Lorado

A Saskatchewan Cold War Uranium Mine and Custom Mill

LAURIER L. SCHRAMM

PATTY OGILVIE-EVANS

and

IAN WILSON

The Authors:
Dr. Laurier L. Schramm, Patty Ogilvie-Evans, and Ian Wilson
Saskatchewan Research Council
125 – 15 Innovation Blvd.
Saskatoon, Canada, S7N 2X8

This book has been carefully produced. Nevertheless, the author and publisher do not warrant the information contained in this book to be free of errors. Readers are advised to bear in mind that the statements, data, illustrations, procedural details, or other items may inadvertently be inaccurate.

The views, opinions, assumptions, and estimates used or expressed in this book are those of the authors and do not necessarily reflect the official policy or position of the Saskatchewan Research Council.

Print ISBN: 978-0-9958081-6-4
ePub ISBN: 978-0-9958081-7-1

DEDICATION

All proceeds from the sales of this book will go to the Saskatchewan Research Council's Technology in Action Fund - a perpetual memorial fund established to help the people of Saskatchewan develop their province as a highly skilled, fair, desirable and compassionate society with a secure environment through research, development and the transfer of innovative scientific and technological solutions, applications and services.

CONTENTS

PREFACE

The Lorado mine, mill, and associated campsites were built in a remote location in northern Saskatchewan, close to Uranium City and just north of Lake Athabasca, the 22nd largest lake in the world.

The Lorado deposit was discovered in 1950 and explored and developed over the following six years, through which the mine was established. The mill was constructed next, and both mine and mill were fully operational by May 1957. A unique feature at Lorado was their mill, which was designed from the beginning to be able to receive and process ores from smaller, neighbouring mines. In addition to being necessary for Lorado's mill to be economic, having a custom mill in the region enabled the development of smaller mines that would otherwise not have succeeded, including Cayzor Athabaska, Black Bay, St. Michael, National Explorations, Lake Cinch, and Rix-Athabasca. The processing of its own ore plus that of other local mines made the Lorado mill the third largest producer of uranium (yellowcake) concentrate in Saskatchewan, and one of the top five producers in Canada during the Cold War era.

By 1960, the market for uranium had crashed, Lorado could no longer sell its uranium concentrate, and both mine and mill were closed. During its relatively short operating life Lorado did, however, play a significant role in helping Canada become one of the largest uranium producers in the world.

The Lorado mine produced about 104 thousand tonnes of uranium ore (not including the uranium ores obtained from other mines) grading about 0.2% (as U_3O_8), and about 54 thousand tonnes of pyrite for the mill. The Lorado mill processed about 469 thousand tonnes of uranium ore (including the uranium ores obtained from other mines) yielding about 1,210 tonnes of uranium (as U_3O_8).

In addition to uranium, the Lorado mine produced a sizeable quantity of waste rock and about 500,000 m³ of tailings covering an area of about 14 ha (35 acres). The tailings, being both acid-bearing and acid-generating, entered into nearby Nero Lake, making it virtually uninhabitable and a threat to other nearby water bodies.

Following closure in 1960, the Lorado site was essentially mothballed and it remained abandoned and deteriorating for the next twenty years, with little remediation being done. The site owners eventually cleaned up most of the mine infrastructure in 1982, and the mill buildings in 1990. Another sixteen years would pass before the government of Saskatchewan contracted the management of the rest of the remediation to the Saskatchewan Research Council (SRC). At the time of writing this book essentially all of the Lorado sites' remediation had been completed, with active monitoring in progress subsequent to ultimately releasing the sites into a long-term management and monitoring program.

ACKNOWLEDGMENTS

Thanks to Ann Marie Schramm, Larry Evans, William Schramm, and Dr. Joe Muldoon for reading and commenting on drafts of this book.

Many thanks also to George Bihun, Saskatchewan Ministry of Environment, for his help finding historical documents and photographs, and to David Thomas and Cory Hughes (Saskatchewan Ministry of Energy and Resources), and Dwayne Pattison (SRC) for supplying additional information on Lorado.

Even in the modern electronic and Internet age, there remains a need for major research libraries with substantive collections of scientific, technical, and engineering books and periodicals. In the preparation of this book, our work was greatly assisted by the collections of the libraries of the University of British Columbia, University of Saskatchewan, University of Regina, University of Calgary, and the University of Toronto.

1 THE URANIUM AGE

1.1 Uranium Exploration – When it All Began (1789-1937).

The mineral pitchblende was first identified and named in 1727 as the rock that glowed[1], at St. Joachimsthal in what is now Czechoslovakia [1]. This was before uranium itself was discovered by German scientist Martin (W.H.) Klaproth in 1789. At first, pitchblende was simply a curious by-product of silver mining, and it seems to have been known in various parts of Europe by the late 1700s. It was later found to be a uranium oxide mineral (also known as uraninite) whose average chemical formula is U_3O_8. Some limited mining of pitchblende for use in colouring glass and porcelain [2] took place in Europe in the early 1800s (and probably in earlier centuries as well), but there was relatively little interest in the mineral until the discovery of X-rays by German scientist Wilhelm Röntgen in 1895, radioactivity by French scientist Henri Becquerel in 1896, and radium in 1898 by French scientists Marie and Pierre Curie (see Reference [3]). Radiation therapy for various diseases, particularly cancer, emerged shortly after the discovery of X-rays, and radium therapy became an even more popular method for radiation treatments beginning in the early 1900s. At about the same time radium came into industrial use as a luminous coating for glow-in-the-dark products, particularly instrument panels, watches, and clocks.

In the United States (U.S.), uranium ore was discovered in 1871 in gold mines near Central City, Colorado and later at the Colorado Plateau of Utah and Colorado. These areas were actively mined for their vanadium and/or radium contents in the late 1800s and early 1900s [4,5].

---

[1] The glow was probably due to radiation from uranium and radium causing zinc sulfide, which is also found in this particular pitchblende mineral, to phosphoresce [1].

Industrial quantities of uranium, in the form of pitchblende, were first discovered at Shinkolobwe in the Belgian Congo[2] in 1915, as brightly coloured "queer stone[s]" [6]. There wasn't much industrial interest in uranium at that time but there was a market for radium[3], which is commonly associated with pitchblende. The Belgian company Union Minière du Haut Katanga mined the Shinkolobwe deposit for its radium, beginning in 1921 and continued for nearly two decades [1,6]. In the early years of radium therapy 100 mg of radium salt sold for about $12,000 (in 1918 U.S. dollars) [7]. When the market for radium severely weakened near the end of the 1930s, the Shinkolobwe mine was closed.

Canada's uranium story had begun somewhat earlier, when in 1900 two scientists from the Geological Survey of Canada, Mackintosh Bell and Charles Camsell, noticed brightly coloured "*lilac stain of cobalt*" on rocks along the shore of Great Bear Lake, Northwest Territories, [6,8]. Bell and Camsell didn't pursue their observations, but they did write a report on their findings. Nearly thirty years later prospector Gilbert LaBine studied their report as part of his research for upcoming prospecting expeditions. In the 1920's Gilbert and his older brother Charles LaBine had opened the Eldorado Mine in Manitoba, and they formed the Eldorado Gold Mines Ltd. in 1926 [9,10]. When the Eldorado Mine played-out, Gilbert LaBine returned to prospecting and began searching near Great Bear Lake [11]. In 1930, LaBine and his prospecting partner Charles St. Paul found Bell and Camsell's cobalt stains and discovered deposits of silver, cobalt, and pitchblende [6,8,9,12]. To develop this find they established a new Eldorado Mine near what would later become Port Radium[4] (see Figure 1.1). They began some mining in 1931/32 and brought the mine and a mill into continuous production in 1933 [8,10].

Uranium-bearing rock (pitchblende) was first discovered in Saskatchewan in 1935 at the Nicholson copper prospect on the north shore of the Athabasca [14] and also in several other locations near the north shore of the lake between 1935 and 1936 [8,15-17].

---

[2] The Belgian Congo achieved independence in 1960, and since then has been the Democratic Republic of the Congo.

[3] At this time radium was in demand for use in cancer treatments.

[4] There were several name changes, first it was called Great Bear, then Cameron's Point in 1932 [1]. By 1933, with prospecting and mining on the increase in this area, the Canadian Government established a town named Cameron Bay, about 11 km south-east of present-day Port Radium. The government facilities in Cameron Bay were renamed Port Radium in 1936 and eventually the whole community came to be referred-to as Port Radium.

Figure 1.1. Illustration of the location of Canada's first uranium mine - the new Eldorado mine and mill – near **Port Radium, east of Great Bear Lake (1)**. Also shown is the location of Eldorado's uranium refinery at **Port Hope on the north shore of Lake Ontario (2)**. The map itself approximates Canada as it was in the 1930s, and was drawn based on the *"Territorial Evolution, 1927"* map in the *Atlas of Canada*, 6th Ed. [13].

None of these were followed-up with significant additional exploration and discoveries until 1944 when Eldorado Mining and Refining staked several claims in the Beaverlodge area (see Figure 1.6, below). Others followed suit in subsequent years. In 1949 uranium exploration in the area had dramatically increased and the abandoned town of Goldfields was revived [15]. Although thousands of radioactive "surface showings" had been discovered by this time, none but Nicholson would be developed until the beginning of the cold-war in 1951. In addition to developing the new Eldorado Mine and the Port Hope refinery, LaBine continued to explore in central Manitoba where, in 1934, he formed Gunnar Gold Mines, which remained in production for several years [18]. As will be discussed below, a second Gunnar mining company was to come later – in the 1950s.

Demand for radium exceeded supply throughout the 1930s, keeping prices high and making radium mining and milling quite profitable. By 1940, however, the adverse effects of radium on human health had become well known, and its use in most medical treatments, consumer-health-products, and luminescent products had been discontinued. As a result, the radium market collapsed. This coupled with the existence of large inventories on hand caused the new Eldorado Mine to be closed in 1940 (as was the Shinkolobwe mine). The Port Radium mine was also virtually abandoned. However, both the Eldorado mine and refinery were destined to become revitalized only two years later, as a result of the international atomic energy race.

1.2 Nuclear Fission is Discovered – The Atomic Age Begins (1938-1941).

In 1938, two German chemists, Otto Hahn and Fritz Strassmann discovered that when they bombarded uranium nuclei the nuclei split apart, yielding two approximately equal fragments of a lighter element, barium, however, the mass of the fragments totalled to less than that of the original uranium [3,19]. Seeking to explain this phenomenon, two Austrian physicists, Otto Frisch and Lise Meitner developed a very important theory. They theorized that the nucleus of a uranium atom, when struck by neutrons, could be split into pieces – the smaller atoms observed by Hahn and Strassmann – plus neutrons, and a huge amount of energy. This process became known as fission. They published their theories in 1939 [20].

Subsequently, Hahn, Meitner, and other physicists realized that such fission actually occurred for a specific isotope of uranium (U-235) that was normally present only in very low concentrations, typically less than one percent. However, if enough U-235 atoms were packed closely enough together, then the neutrons released by one atom could cause the breaking (or fission) of several other uranium atoms, which in turn would split apart releasing more neutrons, and so on, creating a very fast chain reaction and releasing an extraordinary amount of energy [3,21]. The potential to create such huge quantities of energy from very small amounts of material was both very exciting and very frightening.

While excited about the potential for a new source of almost unlimited energy, Frisch and Meitner immediately realized that such power could cause harm as well. Frisch wrote to the British government warning that a small piece of uranium could *produce a temperature comparable to that of the interior of the sun. The blast from such an explosion would destroy life in a wide area ... probably cover[ing] the center of a big city*" [22]. Scientists in other countries, particularly in Germany, Russia, Canada [3,23][5], and the U.S. came to similar conclusions and thus began a race to find and obtain uranium and to try to develop atomic weapons.

It had been estimated that the critical mass needed to enable the chain reaction in an atomic bomb would be about 10 kg (22 lb) of U-235, although the first-ever atomic bomb actually contained 64 kg (141 lb) [22]. Countries involved in the atomic weapons race naturally desired to stockpile as much uranium as possible as a strategic resource for their own use. For a while it was thought that uranium was somewhat rare, with only three substantial deposits being known: Shinkolobwe in the Belgian Congo, Port Radium in Canada (see Figure 1.2), and St. Joachimsthal in a part of Czechoslovakia that had recently been annexed by Germany. The onset of the Second World War in 1939 heightened both strategic interests and fears, as it was also thought that the number of nuclear weapons could be limited by acquiring as much of the known uranium reserves as possible.

Figure 1.2. Photograph of Port Radium in the 1930s. (Courtesy: Public Archives of Canada # C-23966).

---

[5] In Canada, for example, Ernest Rutherford noted in 1904 that "the total energy emitted from 1 gram of radium during its changes is about a million times greater than that involved in any known molecular change ... There is thus reason to believe that an enormous store of energy could be obtained from a small quantity of matter" [23].

These factors set the stage for renewed uranium exploration and atomic power developments in Canada and elsewhere. The National Research Council built Canada's first laboratory-scale fission reactor in Ottawa in 1940 using uranium from Port Hope, via the Eldorado mine that had just closed [24].

1.3 Substantial Uranium Resources Needed – The Atomic Age Reaches Canada (1942-1950).

Canadian exploration for uranium surged again beginning in 1942, due to military interest in building an atomic weapon. Eldorado had been selling their stockpiled, by-product uranium to the U.S. government in 1941 and 1942, but demand kept increasing [10]. In hopes of meeting the increased demand the Eldorado Mine was reopened in 1942 [24-26] and contracted to supply uranium to the U.S. Army, but it was clear that additional uranium reserves would also need to be found [18]. The Eldorado refinery at Port Hope had remained in continuous production, mostly producing radium, but by 1942 its focus had shifted to producing mostly uranium [1].

The Canadian government had imposed a ban on public prospecting for, and mining of, any kind of radioactive materials across Canada, but the government itself vigorously pursued these activities in secret [1,27][6].

In 1942 the United Kingdom (U.K.) and Canadian governments launched a joint atomic energy research program based in Canada [1]. By August 1943 the U.S., U.K., and Canada had merged their atomic weapons development programs under a cooperation agreement called *"The Articles of Agreement on Tube Alloys"* ("Tube Alloys" having been the code name for this project) [18,24]. This agreement was reached and signed in Canada at "The Quebec Conference," held at the Citadel in Quebec City. The "Tube Alloy" Project later became part of the Manhattan Project, led by the U.S. but still in cooperation with the U.K. and Canada. The world's first nuclear fission reactors were built in the U.S. and Canada. Part of Canada's role was to develop and build a heavy water reactor for the production of plutonium from uranium [24].

Included in the Manhattan Project was a component aimed at finding and acquiring as much uranium as possible, beginning in about 1942. At first, the uranium for the Manhattan Project came from the Shinkolobwe Mine in the Belgian Congo and from the Eldorado Mine (refined at Port Hope) in Canada. To this was later added uranium produced in the U.S. itself. Several American companies had been mining for vanadium in Colorado and Utah, but their ore actually contained both vanadium and

---

[6] Attempts to keep Canada's uranium mining secret were not entirely successful. In July of 1943 Eldorado received an order for uranium from the USSR [24].

uranium. The uranium in their ore had now become valuable, but to maintain secrecy the U.S. Army publicly maintained that they were only buying vanadium. At this point in time, the largest known deposit of uranium was still at Shinkolobwe, but there was now a strategic reason to search worldwide for additional reserves: a desire on the part of the U.S., U.K., and Canada to control as much as possible of the world's uranium reserves. This was later enhanced by a growing interest in plutonium (made from uranium).

In 1942, the Canadian Government passed legislation reserving to the Crown ownership of all radioactive substances found in the Northwest Territories and Yukon. In order to maintain security, the Canadian government also started purchasing shares in the Eldorado company [10,24]. In 1943 Eldorado Gold Mines Ltd. was renamed Eldorado Mining and Refining Ltd.[7] [1], and in 1944 the Canadian government expropriated all outstanding shares in the company and turned it into a Crown Corporation. Gilbert LaBine remained President of Eldorado through all of this and had shifted the Eldorado Mine and Port Hope refinery focus to the production of uranium rather than radium. This involved converting uranium ore concentrate (yellowcake) into uranium black oxide (an orange-coloured solid comprising about 96% U_3O_8) [28]. For the next several years, this mine and refinery were the only significant new source of uranium in the western world. Also during this period, the Port Hope refinery received and processed ore from the Shinkolobwe mine [10]. Most of the uranium was sent to the U.S. and probably used to produce the first atomic bombs[8], including the first plutonium bomb ("Trinity") detonated at Jornada de Muerto, New Mexico in July 1945 [9].

In 1946 the Canadian government established the Atomic Energy Control Board (AECB) to regulate the uranium industry and essentially all atomic energy activities [24]. In 1947 Canada began using Eldorado to stockpile uranium, in addition to supplying the U.S. [10]. Meanwhile, the Canadian government pursued additional secret uranium exploration and development activities across Canada using the resources of both Eldorado and the Geological Survey of Canada [12,27]. This wave of uranium exploration was aided by the availability of hand-held portable radiation detectors[9] [29,30]. The most significant finding of this relatively new wave

---

[7] Eldorado Mining and Refining Ltd. later became Eldorado Nuclear Ltd.

[8] It has been estimated that about one-sixth of the uranium delivered to the US Manhattan Project came from Canada [5].

[9] Some of the first commercial hand-held radiation detectors include ionization-chamber detectors, such as the 1930s-era "Curtiss Radium Detector" and the 1940s-era "Victoreen Model 247/247A," and Geiger-Müller counters, such as the 1930s-era "Radium Hound" [29,30].

was that the Beaverlodge region north of Lake Athabasca in northern Saskatchewan not only had pitchblende (as had been previously discovered in 1934) but had at least a thousand pitchblende occurrences [1]! Of these, the first staking took place in 1944 [31] and the first large ore body was discovered in 1946 [1].

In 1948, the Second World War-era security restrictions were reduced, and the federal government repealed the ban on prospecting for uranium by the general public [6,12]. In 1949 the Saskatchewan government established and sold prospecting concessions in the Beaverlodge area[10]. This meant that private enterprises were once again allowed to get involved in exploration, mining, and milling. In addition, technology advances had helped to make low-grade (~0.1 %) ore deposits economically viable. Nevertheless, all mined ores and concentrates were still required to be sold to Eldorado or other government-designated agency (and at a government-guaranteed price) [1,8,10,12,32].

Eldorado's official history [1] notes that: *"Between 1948 and 1953 swarms of prospectors roved far and wide in Canada, many of them amateurs with the exaggerated idea that a few clicks of a Geiger counter would make them rich. There is no record of a really worthwhile discovery having been made by such people, but a number of professional and truly knowledgeable prospectors did 'strike it rich.'"*

The renewed exploration activities of 1948 and 1949 catalyzed the finding and development of new uranium deposits and about 45 small- to medium-size mines [33-35] (see Table 1.1). An example is the Madawaska/Faraday Mine near Bancroft, Ontario, which was discovered in 1949 but did not begin operating until 1957 [36]. The first Saskatchewan uranium mine to be developed was the Nicholson Mine, which commenced production in 1949 [16,35,37].

In 1949 the only Western suppliers of uranium were just Eldorado, Nicholson, and Shinkolobwe but by 1950 both the U.S. and South Africa had also become significant producers of uranium. The price for uranium concentrate guaranteed by the Canadian government was increased in 1950 to encourage further exploration and the development of additional new uranium mines and mills [32].

The end of this era was also the beginning of the Cold War era. The pursuit of military and peaceful applications of nuclear energy, driven by both hope and fear, rekindled demand for uranium. By the end of the 1940s Russia was receiving all of the uranium ore produced by the St. Joachimsthal mine in Czechoslovakia, and by the Schlema mine in East Germany, and it had become known that Russia had successfully test-exploded an underground atomic bomb.

---

[10] These were mostly 65 km^2 (25 mi^2) areas, sold for about $50,000 each [80].

Table 1.1. Examples of Canadian Cold War-Era Uranium Mines.
(Sources: [16,17,33-36,38-40]).

Mine	Discovery	Mean Grade (% U)	Producing Years	Yield (tonnes U_3O_8)
Beaverlodge, SK (Eldorado Ace-Fay-Verna Mines and Mill)	1946	0.24	1953 – 1982	~20,400 [36,38]
Cayzor Athabasca Mine, SK	~1953	0.33	1954-1960	221 [16]
Cinch Lake Mine, SK (later Lake Cinch)	1948	0.20	1955-1960	336 [16]
Eldorado – Dubyna Mine, SK	1947	0.22	1978-1982	192 [16]
Eldorado – Eagle Mine, SK	~1946	Erratic	1950-1951	100 [16]
Eldorado – Fish Hook Mine, SK	1945	0.22	1957-1960	18 [16]
Eldorado – Hab Mine, SK	1958	0.43	1972-1976	900 [16]
Eldorado – Martin Lake, SK	1946	Erratic	1948-1954	13 [16]
Eldorado Port Radium Mine and Mill, NWT	1930		1930–1940; 1942-1960	
Gunnar Mine and Mill, SK	1952	0.18	1955 – 1963	8,133 [16]
Lacnor Mine, ON	1953		1957-1960	2.7 million
Lorado Uranium Mine and Mill, SK	~1953	~0.2	1956-1960	105 [16]
Madawaska/Faraday Mine, ON	1949		1957-1964; 1975-1982	4,305
National Explorations - Keiller , Pat Mines, SK	1951	~0.5-0.8	1954-1958	35 [16]
Nesbitt-Labine Uranium – Eagle, ABC Mines, SK	1950, 1952	0.15-0.24	1952-1956	27 [16]
Nicholson Mine, SK	1935	0.3 – 0.5	1949-1959	48 [16]
Pronto Mine and Mill, ON	1953		1955-1960	2.1 million
Rayrock Mine, NWT	1948		1957-1959	207 [37]
Rix Athabasca – Leonard Mine, SK	1951	~0.2	1955-1960	91 [16]
Rix Athabasca – Smitty Mine, SK	~1949		1952-1960	514 [16]
Uranium Ridges Mine, SK	1950	0.53-0.75	1958-1959	12 [16]

Canada had developed a series of research reactors during this era. The Zero-Energy Experimental Pile (ZEEP) Reactor was Canada's first nuclear reactor and the world's first non-U.S. reactor. It operated from 1945 to 1970 and was used to produce plutonium and uranium-233. Canada's second nuclear reactor, the National Research Experimental (NRX) Reactor, commenced operation in 1947 and remained in service until 1993. Meanwhile, Eldorado had begun to sell cobalt-60, and by 1951 Eldorado, two Canadian university groups (in Saskatchewan and Ontario), and groups in the U.S. had developed cobalt-60 medical devices (called "cobalt bombs") to provide focused gamma rays[11] for radiation treatment of cancer [10,41]. By the early 1950s, Canada had become the world's largest supplier of medical isotopes [42].

1.4 The Cold War-Era Uranium Mines (1951 – 1967).

With the beginning of the cold war, the U.S. decided to continue its nuclear program, including expanding their nuclear arsenal and conducting research and development (R&D) aimed at developing a hydrogen bomb. These activities increased the demand for uranium from Canadian and U.S. mines. Russia had also decided to continue with its nuclear program, drawing uranium from Schlema in Germany and St. Joachimsthal in Czechoslovakia. The early 1950s also saw Britain independently continue its nuclear program, drawing uranium from Portugal, the Belgian Congo, and South Africa. France launched a nuclear weapons development program as well. In Canada at this time, the Port Hope refinery became exclusively focused on producing uranium [1].

Beyond uranium-security and weapons programs, another factor contributing to the demand for uranium was the emergence of nuclear power programs[12]. The United States had started a nuclear power program in the 1940s, and the first electric-power generating nuclear reactor, EBR-I[13], was built in Idaho and started-up in December 1951 [22,43]. For its part, Canada had launched Atomic Energy of Canada Ltd. (AECL, a Crown Corporation) in 1952 and a nuclear power program in 1955, producing the nuclear power demonstration (NPD[14]) reactor, which was built in Ontario

---

[11] Gamma rays are electromagnetic waves of very high energy (and very short wavelength). Gamma rays are one of the kinds of radiation that can be produced by radioactive atoms as they decay.

[12] Industrial applications were also being developed, such as in nuclear density gauges and nuclear thickness gauges, but these did not significantly affect overall uranium demand.

[13] The EBR-I (Experimental Breeder Reactor I) produced 200kW of electricity and was operated from 1951 until decommissioning in 1964.

[14] NPD was the fore-runner of the Canada deuterium uranium (CANDU) power reactors.

and started-up in June 1962 [10,44]. The emergence of these nuclear power programs contributed to governments' desire to build uranium reserves, while regulatory and incentive changes by the Canadian and U.S. governments triggered uranium exploration rushes in both countries [1,4,10].

The regulated (Canadian) price for uranium was increased in 1950 and again in 1951 to stimulate exploration and to ensure that the Beaverlodge and other mines could proceed [32]. As a result, the Beaverlodge mine did proceed (beginning operations in 1953), as did a number of other, smaller mines [1,10]. In response to the "tent cities' that had begun to spring up around the individual mine sites, the Saskatchewan government established the community of Uranium City in 1951 with the aim of serving the entire region (see references [39,45-47]). Then, in 1952, the Saskatchewan government changed regulations making it even more attractive for prospectors to explore and stake claims in the Lake Athabasca region (for which the Canadian Government would have exclusive purchase rights for all mined uranium) and to support them [46]. Figure 1.3 shows an example of the provincial government's accompanying advertising campaign.

The increased uranium price, coupled with the regulatory and infrastructure changes, made it attractive for more companies, and even amateurs to prospect for uranium, triggering a massive uranium exploration and claim-staking rush [4,46,48-52] that helped Canada maintain its international position as a uranium producer[15] (see also Figure 1.4).

A *Northern Miner* headline proclaimed "Uranium - Canada Maintains Place in Frantic World Production Race" [53] and a *Precambrian* article noted that *"Uranium deposits, it seemed, began to appear everywhere"* [54]. The Saskatchewan uranium rush even caught the attention of broadly circulated magazines like *Maclean's* and *Life* [55-58] and made headline news as far away as Australia [59-62] (see Figure 1.5). A television documentary film, *"The Birth of a Great Uranium Area,"* was made in 1953, illustrating the processes of uranium prospecting, drilling, and mining in the area [63]. By the fall of 1954, the government announced that 50 to 60 companies were actively engaged in uranium exploration, development, mining and/or processing in Northern Saskatchewan [64]. Uranium City itself grew to nearly 5,000 people (the size for which it had originally been designed [47]). Another television documentary film, *"The Road to Uranium,"* was made in 1957, illustrating life in Uranium City at its peak of 5,000 people, and showing the operations at the Eldorado mines and mill [65].

---

[15] At about the same time, related developments in the United States triggered a uranium rush there as well [4].

In SASKATCHEWAN

The Spotlight is on

URANIUM

On the north shore of Lake Athabaska in northern Saskatchewan, finds have been discovered which show promise of being the richest uranium source on the North American continent.

URANIUM "HOT SPOT"

Figure 1.3. Excerpt from a 1952 Saskatchewan Government advertisement promoting its "Hot Spot" for uranium exploration (*The Northern Miner*, 1952, *Nov. 27*, p. 51).

The increased level of exploration activity driven by the Cold War led to several significant Canadian uranium discoveries[16] beyond those in Saskatchewan, including deposits around the Bancroft, Ontario, area in the early 1950s (including the Madawaska/Faraday Mine mentioned earlier), and the first discovery in the Elliot Lake / Blind River, Ontario, region in 1953. This led to 12 mines including the Lacnor, Milliken, Nordic (Algom Nordic), Panel, Pronto, Quirke (Algom Quirke), Stanleigh, Stanrock, Can-Met, and Denison Mines [6,27,40] - see Table 1.1 and Figure 1.6.

---

[16] The search for additional uranium resources was not restricted to Canada, of course, and significant new deposits were found elsewhere during this period, such as at Mi Vida in Colorado in 1953.

Figure 1.4. Prospecting with a scintillation counter near Uranium City, *circa*. 1952. Courtesy of the University of British Columbia, Rare Books and Special Collections (Northern Miner fonds).

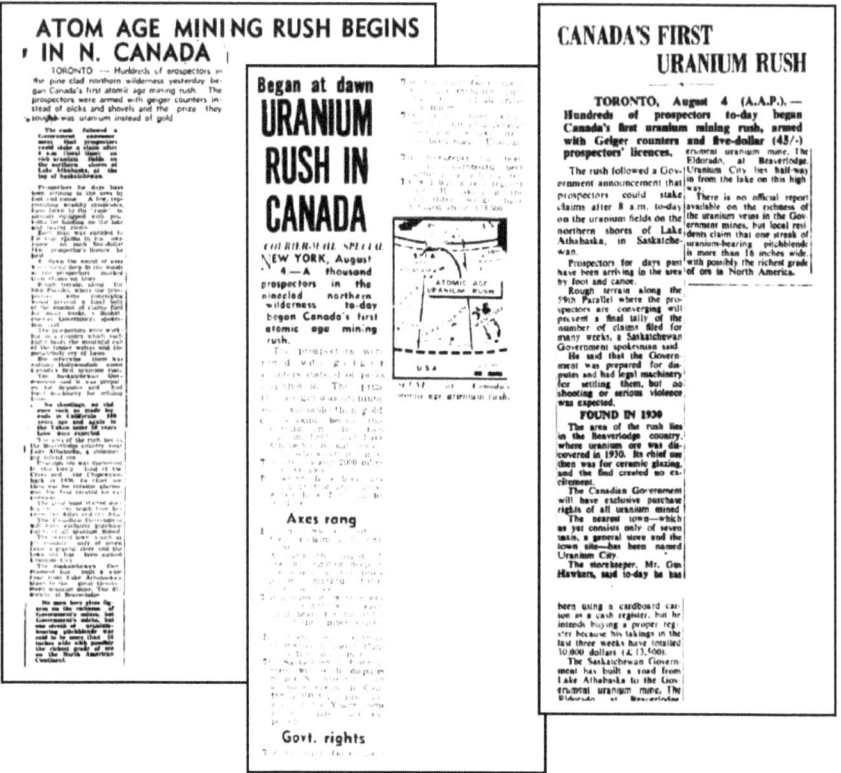

Figure 1.5. Illustration of Australian newspaper headlines reporting the 1952 Saskatchewan uranium exploration boom [59-61].

Prior to World War II the cutoff grade for commercial production of uranium was approximately 0.5% (as U_3O_8), however, the ability to apply more sophisticated processing technologies, and to work at larger scales of operation, brought the cutoff grade down to as low as 0.1 % by the mid-1950s [6]. This change enabled many new mines to be developed in Canada and, again by the mid-1950s, several practical deposits had been identified [27] although some were never developed or did not last very long.

For example:

- In 1949 uranium mineralization was discovered to be associated with silver and lead minerals in the Rexspar deposit in southern British Columbia. The deposit was further evaluated throughout the 1950s (and again in the 1960s and '70s) but was never brought into production [66].
- The Rayrock Mine near Great Slave Lake, Northwest Territories (see Figure 1.6) was discovered in 1948. Production at Rayrock began in 1957, but the mine only lasted for two years [1,36]. The mine and its small town-site were both abandoned in 1959 when its reserves were depleted [1,36].

There had been pitchblende prospecting and staking in Northern Saskatchewan since the 1930s and 1940s, but by 1952 there was enough uranium prospecting activity that the Saskatchewan government established the town of Uranium City, as noted above, to provide services to the mines in the Beaverlodge area north of Lake Athabasca (see Figures 1.6 and 1.9). Several Saskatchewan discoveries followed that of Beaverlodge, including Radiore nearby, Rix Athabasca on Black Bay, and Nesbitt-Labine near Uranium City [10]. Gilbert LaBine and his son Joseph were part of the renewed prospecting in this region, and they found several deposits in the same general area in 1952 [6,9].

In July 1952 LaBine, Albert Zemel, and Walter Blair found and staked a uranium deposit at the southern tip of the Crackingstone Peninsula on Lake Athabasca [1,67] (see Figure 1.7). This became the site of the Gunnar Mine, which was the richest uranium strike in Canada at the time and would become the first large private uranium mine of the era [10,58]. When the Gunnar Mine and mill site opened in the fall of 1955, it doubled Canada's uranium production capacity [68-70] (Figure 1.7).

Through Eldorado Nuclear the LaBines began mining at their Beaverlodge site[17] in 1953, where they established a dedicated mill and also a small community named Eldorado, all located about 7 km east of Uranium City [27,38]. This site, which included the Ace, Fay, and Verna mines[18], operated until 1982 and was the highest producer of the 16 Beaverlodge area mines of the Atomic Age and the Cold War Eras (Table 1.1; [69]).

---

[17] The Beaverlodge Mine has sometimes been referred-to as the Eldorado Mine, although as noted above there were two previous Eldorado Mines, one in Ontario and an earlier one in Manitoba.

[18] An illustration of the early days of establishing Eldorado's Ace and Fay mines is given in a TMC documentary [63].

Figure 1.6. Illustration of the locations of some of Canada's cold war era uranium mines: Eldorado Mine (1), Bancroft-area mines (2), and Elliot Lake - Blind River area mines (3), Rayrock Mine (4), Beaverlodge-area mines (5). The map itself approximates Canada as it was in the 1950s, and was drawn based on the "*Territorial Evolution, 1949*" map in the *Atlas of Canada*, 6th Ed. [13].

The Lorado mine and mill officially opened in 1957 (although some mining had been conducted earlier) and operated until 1960 [36] (see Figures 1.8, 1.9). Lorado was the only other mine to have its own dedicated mill (along with Beaverlodge and Gunnar), and the Lorado mill operated from 1957 to 1961. In addition to processing the mined ore from their own mines, the Lorado and Beaverlodge mills also processed ore from smaller mines in the region, including the Cayzor, Rix Leonard, and Cinch Lake Mines, and also ore from "surface miners" who picked over surface showings and waste rock piles that were uneconomic for the mining companies to handle [36,67].

Figure 1.7. The Gunnar open pit mine in 1959 [71].

In total, more than 1,000 pitchblende occurrences were discovered in Saskatchewan's Beaverlodge district, however, only 16 mines of significant size were actually brought into production between 1953 and 1982 (see Table 1.1) [16]. By 1960 LaBine estimated that the Beaverlodge district mines had produced close to $300 million worth of uranium (in 1956 dollars) [31].

By 1957 there were 18 operating uranium mines in Canada (see Table 1.1), all of which sold their uranium ore (raw or milled) to Eldorado, which in turn sold the refined uranium to the U.S. [10]. This number peaked at 21 Canadian producing mines in 1958, by which time 11 of the mines plus three mills were operating in the Uranium City area [1,12,36].

Canadian uranium production levels grew throughout the 1950s. In 1956 the "free-world" production of uranium was close to 13,000 tonnes [6]. Most of the production came from the U.S., Belgian Congo, and Canada (see Table 1.2). By 1958 half of the world's production of about 27,200 tonnes was estimated to have come from "Arctic Canada" meaning the Port Radium Eldorado Mine [6,27]. By 1959 uranium was Canada's number one mineral export (ahead of aluminum, iron, and nickel) with 23 mines and 19 mills in operation [27]. Of the 19 mills, 11 were in the Elliot Lake, Ontario area, three near Bancroft, Ontario, three in northern Saskatchewan, and two in the Northwest Territories [27].

Figure 1.8. The Lorado uranium mill in 1958 [72].

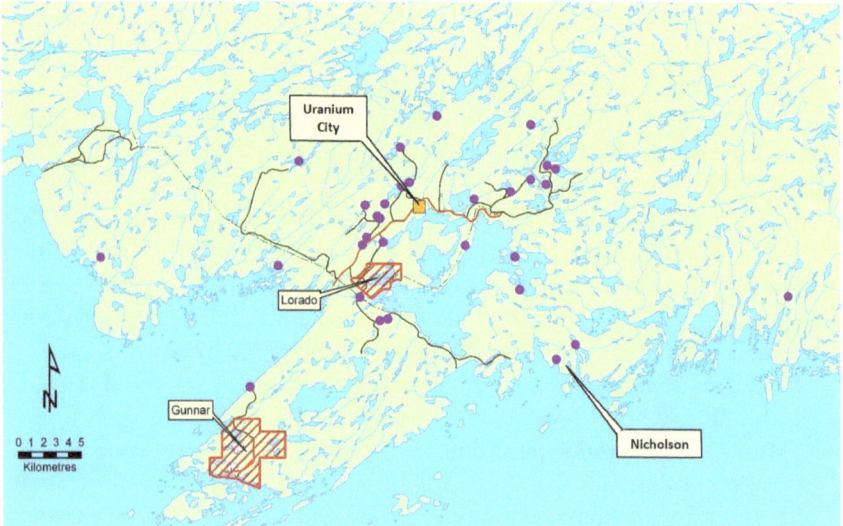

Figure 1.9. Illustration of the locations of three of Saskatchewan's cold war era uranium mines: the Gunnar, Lorado, and Nicholson Mines.

Table 1.2. World uranium reserves in 1957.
(Conversions to S.I. units are approximate. Based on data in Reference [6].)

Country	Production in 1956 (tonnes U_3O_8)	Uranium Reserves (millions of tonnes U_3O_8)	Average Ore Grade (mass%)
South Africa	3,992	998	0.03
Canada	2,994	204	0.10
United States	5,443	54	0.24
Total	12,429	1,256	

By the late 1950s, radiation safety was becoming better understood. Although by 1940 the adverse effects of radium on human health had become well known and its use in most medical treatments and consumer products had been discontinued, the health effects of low doses of radiation were not yet well known. The concepts of safe working levels and safe cumulative (annual) exposure levels emerged in the 1950s, although uranium mining in the 1950s was still considered "safe" (as far as radiation hazards were concerned) [24,73]. In 1959 Gordon Churchill, the federal Minister responsible confidently stated in the House of Commons that "... *there are no special hazards attached to the mining of uranium that differ from other mining activities*" and "... *there is no radiation hazard in the processing operations*" [24]. Nevertheless, in 1960 the AECB created regulations dealing with radiation safety [24]. The 1960 AECB regulations defined for the first time the concept of an "*atomic energy worker*" and the maximum amounts of ionizing radiation to which such a worker could be allowed to become exposed[19].

Most (about 90%) of the Canadian uranium produced in the 1950s and early 1960s was sold to the U.S. Atomic Energy Commission, with most of the remainder being sold to the U.K. Atomic Energy Authority [1,8,12]. The Rayrock mine closed in 1959. Then the Eldorado Mine closed in 1960, when the uranium ore ran out, although the Eldorado mill was able to continue operating until 1967. Despite depleting ore bodies, by this time

---

[19] These regulations also defined the maximum amounts of ionizing radiation to which a member of the general public could be allowed to become exposed, at 1/10th of the amount for an atomic energy worker.

uranium production was outstripping demand. By 1962-63 the U.S. had more than enough uranium for its needs resulting in the U.S. Atomic Energy Commission reducing its purchases [29]. As a result, the levels of both exploration and mining and milling decreased, and the number of active mines shrunk to only five, the Madawaska/Faraday, Milliken, and Nordic mines in Ontario, and the Beaverlodge and Gunnar mines in Saskatchewan [12,27]. A 1966 Saskatchewan Department of Mineral Resources report concluded that *"uranium was a glut on the market, and without markets incentive is lacking to search for and develop new mines"* [12]. For a while, the Canadian government supported the uranium industry with a stockpiling program, however, this only lasted until 1974 [27].

Although there was an oversupply of uranium in the markets, the late 1960s were also characterized by much eager anticipation of nuclear power and clean energy. As already noted, the first power-generating nuclear reactor, EBR-I, had begun operating in Idaho, but not many other power reactors had yet been built. Nevertheless, with forecasts that the world's known reserves of uranium[20] could be quickly used-up if many nuclear power plants were constructed [12], prospecting for new deposits increased again, and once again Saskatchewan was *"now one of Canada's busiest [uranium] prospecting areas"* [12].

Uranium production in the Bancroft and Beaverlodge areas ended in 1982, and in the Elliot Lake area in 1996 [27]. Thus ended the second era of Canadian uranium production. The first two eras of Canadian uranium production helped put Canada on the world stage from an industrial point of view, and these eras are rich in stories. One of these is the story of the Lorado Mine, which is described beginning in Chapter 2[21].

The Beaverlodge area mines and smaller developments of the Atomic Age and the Cold War Eras left behind an unfortunate legacy in that they were simply abandoned without much or any cleanup, and frequently without significantly closing-off the various mine shafts, adits, and raises. A 2006 Saskatchewan Environment and Resource Management report identified 45 such abandoned mines in the immediate Uranium City area alone [41]. These aspects are described further, in the context of the Lorado operations, in Chapters 7 and 8.

There was a significant lull in Canada's uranium production beginning with the end of the cold-war era in around 1967. This was partly due to market conditions, as described above, and also partly due to changes in the

---

[20] In 1964 the known world reserves of uranium were 430,000 tonnes, 40% of which was located in Canada (Saskatchewan Department of Mineral Resources, 1966).

[21] Two other significant Saskatchewan Cold-War uranium mine stories are those of the Gunnar Mine, which is described in reference [70], and the Nicholson Mine, which is described in reference [36].

political and regulatory environment. Although the Atomic Age and the Cold War Eras (1938 through 1967) exhibited uranium mine developments across Ontario, the Northwest Territories, and Saskatchewan (see Table 1.1), the modern-era uranium developments in Canada would all take place in Saskatchewan. (Saskatchewan also has one nuclear reactor in operation). The Saskatchewan Research Council has operated a SLOWPOKE 2 research reactor in Saskatoon since 1981 [74].) Between 1982 and about 1992, the Saskatchewan government phased-out uranium mines and mills, however from approximately 1992 onwards the province reversed course and once again became an active supporter and co-developer of uranium mining in Saskatchewan [75]. Whereas the early Saskatchewan developments were in the Beaverlodge area immediately north of Lake Athabasca, the next boom in uranium exploration (in the 1970s) resulted in huge uranium discoveries in the Athabasca Basin area immediately southeast of Lake Athabasca. These deposits were not only large but were of extremely high grade, enabling a huge increase in Canadian uranium production. From approximately 1992 through 2008 Saskatchewan became the uranium capital of the world with the highest productions levels and the largest deposits of the highest grade of uranium on the planet. By 2008 more than ten times as much uranium had been produced from the Athabasca region as was produced during the entire operating history of the Beaverlodge region (322 thousand tonnes vs. 30 thousand tonnes as U_3O_8) [66]. All of Canada's modern-day uranium production comes from mines in northern Saskatchewan (see Figure 1.10) [76] and has been close to 12,900 tonnes of uranium per year from the late 1990s to the present [77].

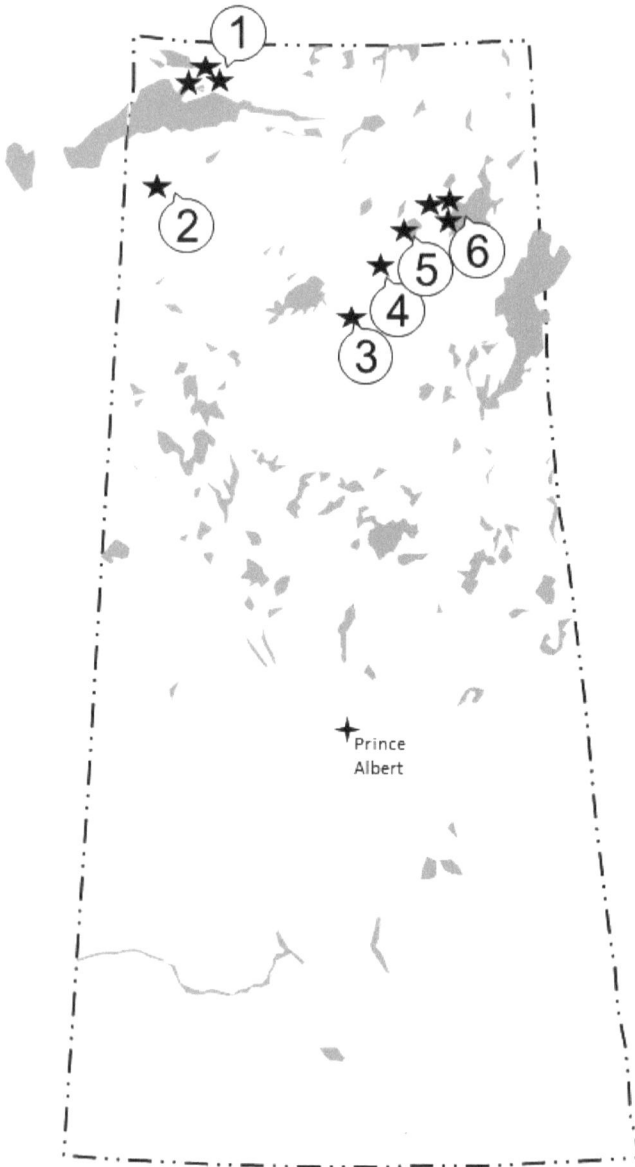

Figure 1.10. Illustration of the locations of Saskatchewan's cold war- and modern-era uranium Mines: Lorado, Gunnar, Beaverlodge, and Nicholson mines (1); Cluff Lake mine (2); Key Lake mine (3); McArthur River mine (4); Cigar Lake mine (5); and Midwest (proposed), McClean Lake, and Rabbit Lake mines (6). The city of Prince Albert is shown for reference. Drawn based on the "*Saskatchewan*" map in the *Atlas of Canada*, 6th Ed. [13].

2 THE BEGINNINGS AT LORADO

2.1 The Discovery and Initial Exploration of the Lorado Site.

As noted in Chapter 1, there had been pitchblende prospecting and staking in Northern Saskatchewan in the 1930s and 1940s. Then, between 1950 and 1952 a "uranium rush" had developed in Saskatchewan.

The forerunner to the Lorado mine was the staking of seven claims, known as the Alco (1-7) group, in 1950, by gold prospector Gunnar Berg[22]. These claims were staked for Noranda Mines Ltd. as part of the former "BB" concession [17,79,80,81].

The general location of the Lorado site is north of Lake Athabasca (Figures 1.6, 1.9, 1.10, 2.1, and 2.2). While this site is "near" Uranium City, it is approximately 10 km to the southwest and is accessible by the main road between Bushell and Uranium City.

A Geological Survey of Canada map showing the broad age groupings of deposits in and around the Lorado site is shown in Figure 2.3, and a surface map showing the seven Alco claim areas is shown in Figure 2.4.

The specific locations of the Lorado mine and mill sites are identified in Figures 2.2 and 2.3.

Noranda's initial exploration on the claims involved nearly 8,000 metres (25,000 feet) of diamond drilling and indicated "*a few scattered but interesting zones of radioactivity*" [15].

---

[22] Gilbert LaBine named the Gunnar Gold Mines Ltd. company (the forerunner of Gunnar Mines Ltd.) after Gunnar Berg [78].

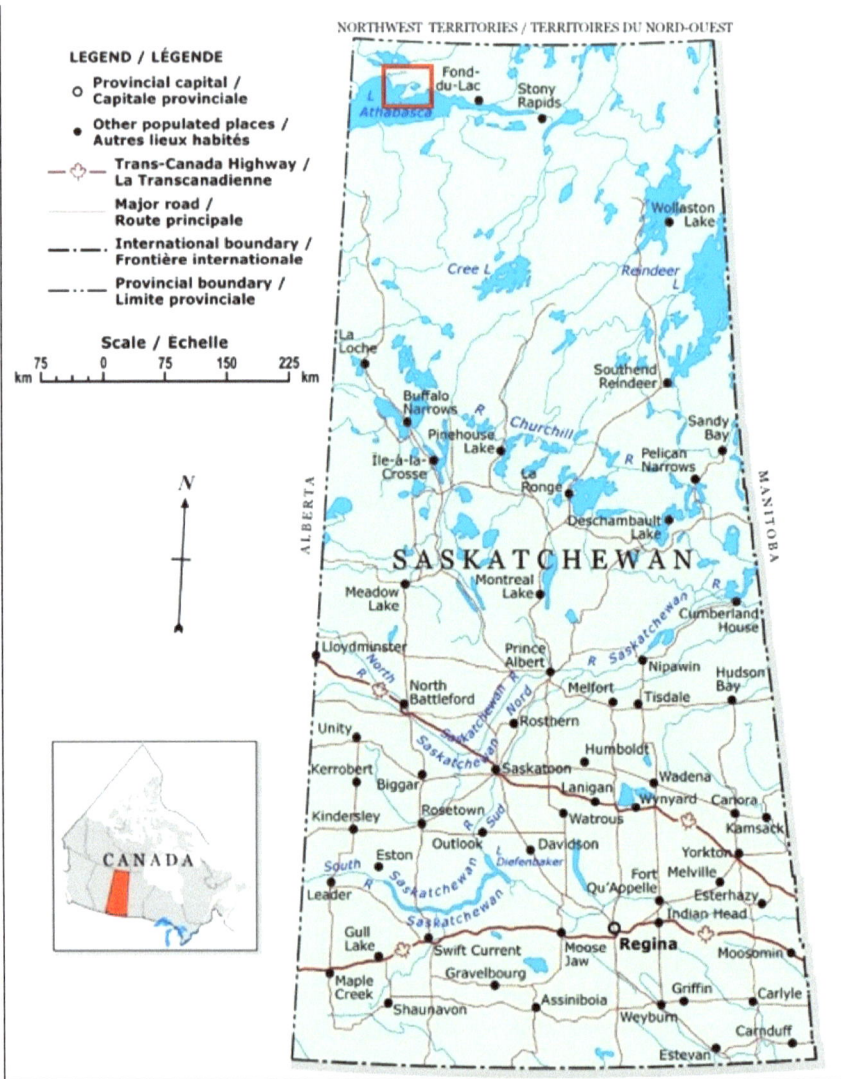

Figure 2.1. Lorado site location (red square in the extreme upper left corner of the map. (Map from Natural Resources Canada, 2001.)

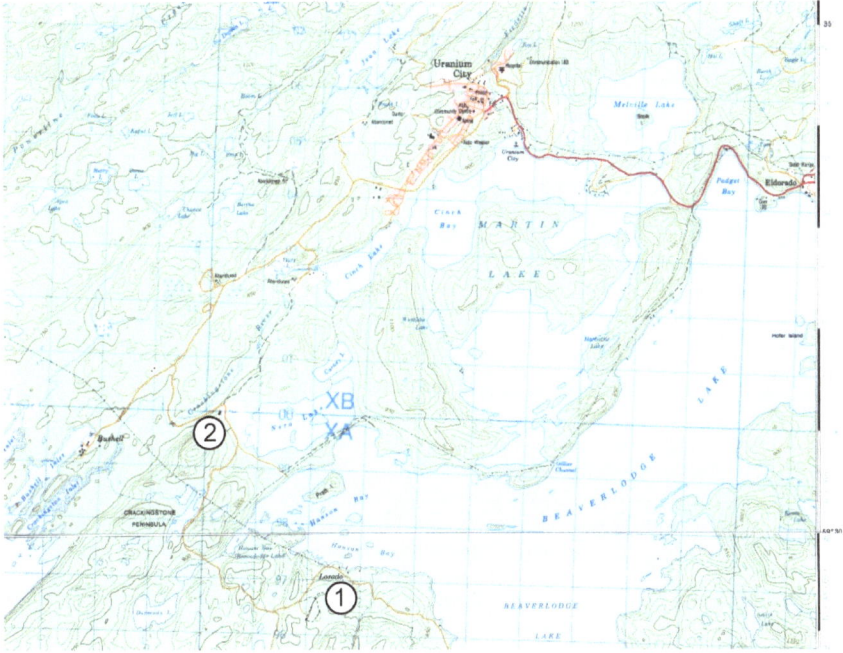

Figure 2.2. National Topographic System (NTS) map excerpt showing the Lorado Mine site (1) and the Lorado Mill site (2). The mine was south of the western edge of Beaverlodge Lake, while the mill was further north, adjacent to the western shore of Nero Lake. Uranium City is shown to the north, and Bushell is shown to the west of the Lorado mill site. Adapted from NTS Maps Uranium City, 74 N/10 and Crackingstone Penninsula 74 N/7 & 74 N/6, Energy, Mines, & Resources Canada, 1988.).

Noranda did not determine their initial exploration results to be sufficiently encouraging, and in 1952, the company sold the property to K. Sherman Oliver, who formed Lorado Uranium Mines Ltd. in April of the same year [79]. Oliver became President and established the company's Head Office in Toronto, with the Mine and Mill Offices in Uranium City [82]. The company immediately set about raising capital (see Figure 2.5).

Lorado commenced new drilling activities on the Alco claims almost immediately [83,84]. In addition to the Alco claims, Lorado conducted exploration and diamond drilling assessments on several other claims in the Beaverlodge Lake area, including Alcan-Geil-Dix, Dot-2-22, and at Viking Lake, unfortunately without getting encouraging results beyond what was found on the Alco claims [80,81,85].

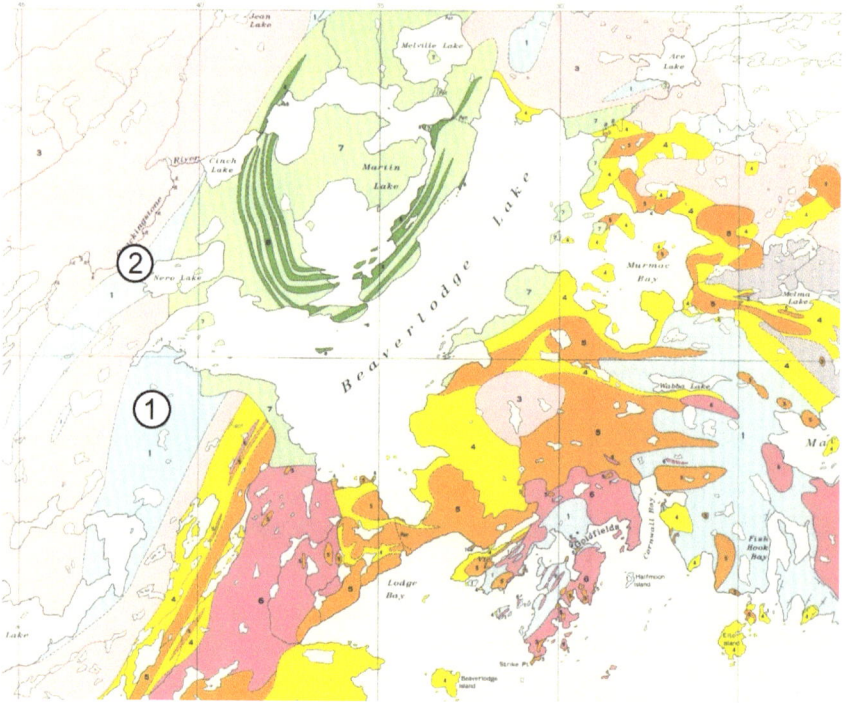

Figure 2.3. 1936 Geological Survey of Canada map showing the Lorado Mine site (1) and the Lorado Mill site (2). The zones near the Lorado mine are all of Archean (Early Precambrian) age and are shown as Tazin Group (blue), Post-Tazin (pink), and Beaverlodge Series (yellow). From the Geological Survey Map 339A, "Goldfields Area," 1936, reference [14].

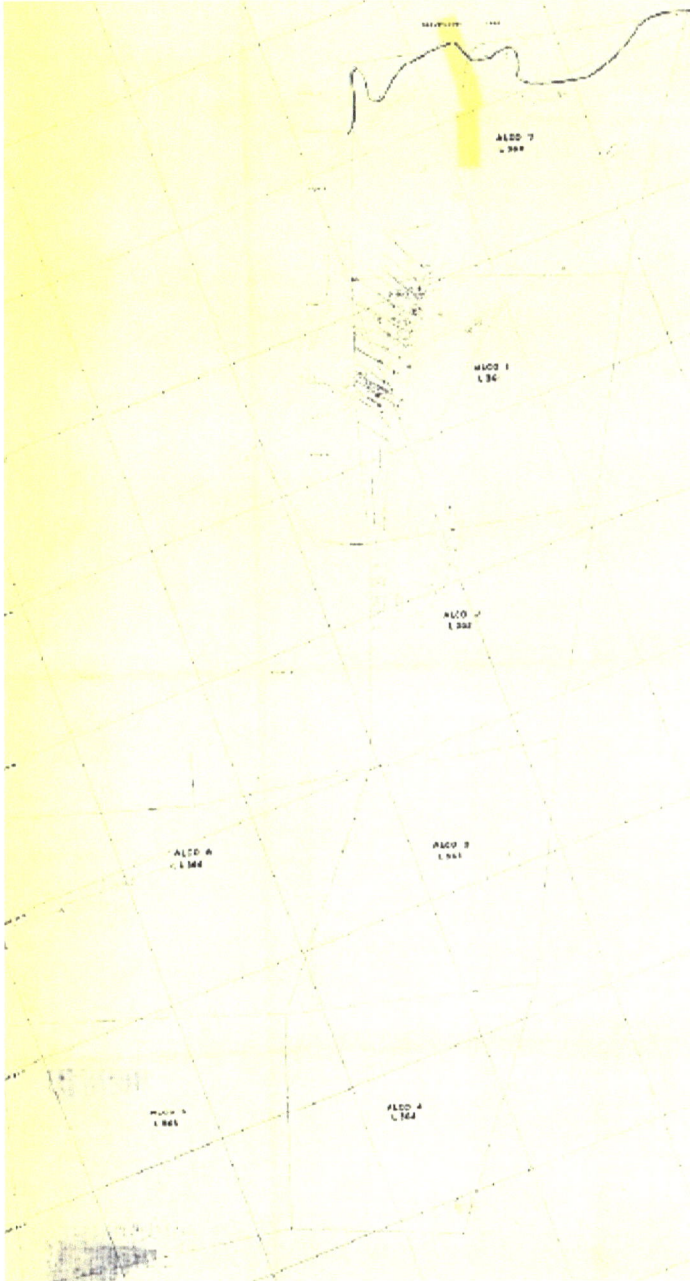

Figure 2.4. Annotated surface map showing the seven Alco claim areas (numbered clockwise from the top right: 7,1,2,3,4,5,6). The mine is shown centred in the upper left of Alco claim 1. (Lorado Uranium Mines Ltd., 1959.)

Figure 2.5. Excerpt from a 1953 newspaper ad promoting sales of Lorado Uranium Mines Ltd. shares (*The Northern Miner*, 1953, May 7, p. 5).

2.2 Early History of the Mine and Mill.

In 1953, a shaft was sunk to the "200-foot level[23]" (actually about 65.5 m, or 215 feet, down) on the most promising part of the claim areas, which was in the northeast corner of Alco Claim #1 (Figures 2.4, 2.6). As continued drilling remained promising in 1953, a second diamond-drilling rig was brought in and news updates hinted at a significant ore body with samples grading in the range of 0.2% uranium oxide, U_3O_8 [86-88] (see Figure 2.7). Drill cores were initially sent to a Winnipeg lab for assaying and later sent to the nearby Pitch-Ore Uranium Mine when that company opened its own onsite chemical assay laboratory [89]. By the fall of 1954, Lorado had its own on-site chemical assay laboratory and had begun to offer assaying services to other uranium companies as well [90].

With help from the Government of Saskatchewan, road access from Bushell and Eldorado to the Lorado site was improved during 1954 to permit all-weather traffic under heavy loads [91]. This, in turn, enabled the commercial-scale developments of mine and mill infrastructure.

Figure 2.6. The Lorado headframe and hoist under construction in 1953. Courtesy of Saskatchewan Archives Board (Photo R-B5412-4).

---

[23] In some media reports this was referred to as the 225-foot level.

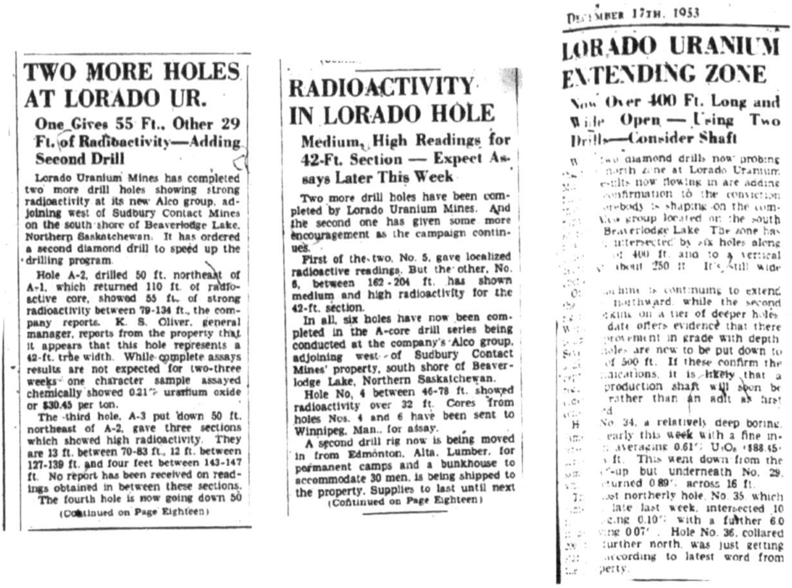

DECEMBER 17TH, 1953

TWO MORE HOLES AT LORADO UR.

One Gives 55 Ft., Other 29 Ft. of Radioactivity—Adding Second Drill

Lorado Uranium Mines has completed two more drill holes showing strong radioactivity at its new Alco group, adjoining west of Sudbury Contact Mines on the south shore of Beaverlodge Lake, Northern Saskatchewan. It has ordered a second diamond drill to speed up the drilling program.

Hole A-2, drilled 50 ft. northeast of A-1, which returned 110 ft. of radioactive core, showed 55 ft. of strong radioactivity between 79-134 ft., the company reports. K. S. Oliver, general manager, reports from the property that it appears that this hole represents a 42-ft. true width. While complete assays results are not expected for two-three weeks one character sample assayed chemically showed 0.21% uranium oxide or $30.45 per ton.

The third hole, A-3 put down 50 ft. northeast of A-2, gave three sections which showed high radioactivity. They are 13 ft. between 70-83 ft., 12 ft. between 127-139 ft. and four feet between 143-147 ft. No report has been received on readings obtained in between these sections.

The fourth hole is now going down 50
(Continued on Page Eighteen)

RADIOACTIVITY IN LORADO HOLE

Medium, High Readings for 42-Ft. Section — Expect Assays Later This Week

Two more drill holes have been completed by Lorado Uranium Mines. And the second one has given some more encouragement as the campaign continues.

First of the two, No. 5, gave localized radioactive readings. But the other, No. 6, between 162-204 ft. has shown medium and high radioactivity for the 42-ft. section.

In all, six holes have now been completed in the A-core drill series being conducted at the company's Alco group, adjoining west of Sudbury Contact Mines' property, south shore of Beaverlodge Lake, Northern Saskatchewan.

Hole No. 4 between 46-78 ft. showed radioactivity over 32 ft. Cores from holes Nos. 4 and 6 have been sent to Winnipeg, Man., for assay.

A second drill rig now is being moved in from Edmonton, Alta. Lumber, for permanent camps and a bunkhouse to accommodate 30 men, is being shipped to the property. Supplies to last until next
(Continued on Page Eighteen)

LORADO URANIUM EXTENDING ZONE

Now Over 400 Ft. Long and Wide Open — Using Two Drills—Consider Shaft

Two diamond drills now probing north zone at Lorado Uranium. Results now flowing in are adding confirmation to the conviction ore body is shaping on the company's Alco group located on the south shore of Beaverlodge Lake. The zone has intersected by six holes along of 400 ft. and to a vertical about 250 ft. It's still wide.

... zone is continuing to extend northward, while the second drill on a tier of deeper holes. holes are now to be put down of 500 ft. If these confirm the indications, it is likely that a production shaft will soon be rather than an adit as first ...

No. 34, a relatively deep boring, early this week with a fine, averaging 0.61% U₃O₈ $88.45 ft. This went down from the up but underneath No. 29, turned 0.89% across 16 ft. ... northerly hole No. 35 which late last week, intersected 10 ... ing 0.10% with a further 6.0 ... ng 0.07'. Hole No. 36, collared further north, was just getting ... according to latest word from ... perty.

Figure 2.7. Examples of 1953 newspaper articles tracking the drilling program at Lorado [86-88].

Through 1954 and 1955, the mine shaft was extended to a depth of 700 feet (about 213 m) with levels at 215, 360, 500, and 600 feet (i.e., about 66, 110, 152, and 183 m) [15,17]. During this same time period, lateral developments were undertaken at the various levels [92].

Additional drilling through 1954 continued to produce encouraging results generating some excitement in the region and in the media (Figure 2.8). In 1955 Oliver summarized the ore body delineation results as "2,500 tons per vertical foot of ore grading 0.20 percent uranium oxide" [91].

Test work on the processibility of the Lorado ore was begun in 1953 and completed in 1955, after which a processing (milling) flow sheet was developed. This formed the basis for the mill to be designed [82,91]. By the fall of 1955, the company already had a mine-site workforce of 110 and was preparing to expand the campsite facilities to accommodate the additional staff that would be needed to build[24] and operate a mill [96].

---

[24] The mill was designed by Kilborn Engineering, and built for Lorado by McNamara Construction Co., the latter using a crew of 235.

PICTURE GROWS AT LORADO

Officials See Single Large Zone—Now Opened For 265 Ft.—Values Exceed Drilling

Recent drifting results at Lorado Uranium Mines continue to open an excellent grade of ore in the main zone, which now looks to be continuous for a length of 265 ft with both ends open. And widths, where slashed, look pretty good.

The main south drive, 100 south drift (see sketch) has intersected underground drill hole No. U-2, where channel sampling has far exceeded the drill hole returns. The hole itself returned a 111.5 ft intersection that averaged 0.14% U₃O₈. While the section of the hole blasted into showed only 0.16% in the drill core, channel sampling of this section returned 0.45% over the drift width $65.25 at $7.25 per lb.

Adding further to officials' conviction that the drift values have been consistently on the low side because of poor core recovery, a 6.0 ft. slash just taken along the drill hole to the east on a section that assayed only 0.65% has returned by channel sampling 0.97%. There is a lot of gummite and pitchblende showing on the walls, K. S Oliver, president and mine manager tells The Northern Miner.

BIG ORE WIDTH AT LORADO

Lorado Uranium Mines has further evidence that its main zone is going to be both wide and of excellent grade. The first crosscut has been completed.

It shows a 120-ft. continuous ore width that averages 0.33% U₃O₈ ($47.85 at $7.25 per lb.). This figure represents both channel sampling of the faces and rib samples cut along the walls and back of the heading, officials point out to The Northern Miner.

The above crosscut, which is the first of a series to be driven across the zone at regular intervals as drifting advances, followed along underground drill hole No. U-1, which returned 0.14% for 111.5 ft. The improvement in results therefore adds to the conviction of the management that the drill values have been consistently low because of poor core recovery.

Drifting south along the main zone has now been resumed.

Drifting north along the main zone continues to open fine ore, with assays on three additional rounds reported since last week's story. These latest rounds assayed 1.05%, 0.24% and 1.31%, respectively, again representing channel samples across the drift width. Last of the above faces has been slashed out to a 100-ft. width, with the slash itself assaying 0.95% U₃O₈.

LORADO DRIFTING LENGTHENS ORE

Main Zone Now Shows 360 Ft. Continuous Ore — Open To North—Driving to South

Drifting on the 220-ft. level at Lorado Uranium Mines' Alco property in the Beaverlodge area now shows a continuous ore length of 360 ft. in what is considered the main zone, K. S. Oliver, president, told The Northern Miner this week. While further work will be required to determine full width and overall grade, both look impressive.

The 105 drift which is being pushed northward (see sketch, in The Northern Miner, Nov. 25) is particularly rich ore. Radiometric assays for the last three faces averaged 3.27% U₃O₈ across 6.7 ft. Individually, they ran 4.94%, 3.12% and 1.76%. A chemical assay to check the first of these returned 6.49% U₃O₈. Both walls and the present face of the heading are described as being in similar material. This drift now shows a continuous ore length of 105 ft. averaging 0.90% across 7.2 ft. It is being continued northward.

To determine the full width of the zone, flat holes are to be put out from the drifts at 50-ft. intervals with stub crosscuts to be driven at 150-ft. intervals. As reported last week the first crosscut in this series showed an ore width of 120 ft. that averaged 0.33% U₃O₈.

Figure 2.8. Examples of 1954 newspaper articles tracking the mine development program at Lorado [93-95].

Beyond uranium mining, it was further decided to mine Lorado's own deposit of pyrite, to provide the raw material for the manufacture of sulphuric acid for the milling process. The company estimated that it had over 2.7 million tonnes (3 million tons) of pyrite ore grading between 30 and 40 percent pyrite [82]. The possibility was contemplated to manufacture additional sulphuric acid for sale to other mills in the area [91].

By 1956, the company had identified significant uranium reserves on the first and second levels but had only found suggestions of ore-grade uranium mineralization at the lower levels [82]. At this stage, it was clear to the company that they could not prove sufficient reserves to obtain a contract with Eldorado Mining and Refining Ltd. at the *"minimum economic rate of 500 tons per day before the [Eldorado] deadline of March 31st, 1956"* [82]. On the other hand, as early as 1955, Lorado management had recognized that several other companies in the area were going to be in a very similar, but worse situation having even smaller reserves than did Lorado [97,98]. As a result, Lorado decided to develop a custom mill that could process both their own mined ores as well as those of the smaller, nearby uranium mines. On this basis, Lorado applied to Eldorado *"for a custom milling franchise and a special price contract to enable operation on a practicable basis"* [98,82].

In 1956, Lorado was one of fifteen companies that received special uranium-purchase contracts from Eldorado that were intended to cover the necessary pre-production costs and capital outlays over the life of the contracts (i.e., five years, ending in March 1962) [83]. Some of these contracts were substantial, amounting to $60 million for Lorado Uranium

Mines and $77 million for Gunnar Mines [82,83]. In Lorado's case, this contract was judged to be sufficient to support *"full scale operation of a 500 ton per day mill"* [82]. By spring 1956, the company had ordered *"all of the major components and equipment for the 500-ton mill and acid plant"* [99], and by the fall of the same year, the mill buildings were visibly taking shape (see Figures 2.14-2.17).

Lorado's 1956 Annual Report notes that *"After reserving sufficient mill capacity to treat [their] ore the remaining capacity was allotted to, and custom milling contracts were entered into with, Cayzor Athabaska Mines Limited, Black Bay Uranium Mines Limited, St. Michael Uranium Mines Limited, and National Explorations Limited. Subsequently, when the amount of the Eldorado contract was increased ..., a custom milling contract was entered into with Lake Cinch Mines Limited"* [82]. With the addition of the Lake Cinch Mines contract, it was decided to increase the mill's design capacity from 500 tons per day to 750 tons per day [82]. Another custom milling contract was signed in 1958, this time with Rix-Athabasca Uranium Mines Ltd. and Lorado also accepted ores from various smaller mines over its producing years [101].

Figure 2.9. Rock sample, with uraninite showing (centre), from the Lorado Mine. (Lorado Uranium Mines Ltd., 1956 [82].

Construction of the mine and mill were enabled through public financing. On August 27, 1956, a syndicate led by McLeod, Young, Wier & Co. issued $9,250,000 worth of sinking fund debentures to mature on March 1, 1962, at par to yield 6% [82,102]. These debentures also carried stock purchase warrants.

The mine itself was located near the southwest corner of Beaverlodge Lake (see Figure 2.2). The mill was located along the same road, close to 6.4 km (4 mi) northwest of the mine, closer to the Bushell end. This placed the mill just to the west side of Nero Lake, a small lake that drains into Beaverlodge Lake, which in turn drains into the Crackingstone River, and from there to Lake Athabasca. It was unusual not to have the mill located adjacent to the mine, although, in Lorado's case, there was a specific reason for this.

As already noted, realizing that they did not have sufficient proven reserves to justify the construction of a large enough mill to be cost-effective, and with the support of Eldorado Mining and Refining, Lorado decided to develop a custom mill that could accept mined ores from neighbouring mines in addition to those from Lorado's own mine [15,82]. With this in mind, a mill site was chosen that would be in a more central location, with respect to the neighbouring mines that they hoped would become customers for their milling services [15,103].

Being in a remote location, and without road access, most construction and other materials had to be brought in by boat, barge, or by air[25] (with an aircraft landing on either Lake Athabasca or Beaverlodge Lake).

Another complication was the long lead times, ranging from six to ten months, required to even obtain major equipment such as ball mills, diesel engines, and filters [82]. Accordingly, supplies for the new mill were ordered early in 1956, in order to take advantage of the water-shipping season, and at the same time, Lorado started sawing logs in their own sawmill, all in order to be ready for use for building construction beginning in the spring of 1957.

The preferred and most economical option for Lorado, and indeed for all the uranium mines in the region, was to use a barge service. Barges were towed by shallow-draft tugboats, such as those of the Hudson's Bay Co. or the Northern Transportation Co. [104,105]. The principal barge route at the time was about 440 km (265 mi), from either Uranium City or nearby Bushell, along Lake Athabasca to the Athabasca River and then to the railhead at Waterways, Alberta, which in turn was served by the Northern Alberta Railway [8,67,106] (Figures 2.10 - 2.12). This provided close to six months of ice-free waterway per year, of which only about four months per year were practical for waterborne transportation given the high winds and

---

[25] Operated mostly by McMurray Air Services, which had a seaplane base at Uranium City.

low water levels that characterized each fall season [31,68,107]. The problems were amply demonstrated when, according to Quiring [108], "*a boat towing eight barges loaded with material for Gunner and Eldorado froze into the ice in 1956.*"

Eldorado found that barging conditions ranged from "*shallow bar ridden rivers to deep and frequently rough lakes,*" and that late summer water levels could fall as low as 50 cm (20 inches) in some areas, restricting the weight of loads that could be carried, and therefore increasing the cost per tonne shipped [109]. Despite these obstacles, transportation by barge continued to be the preferred option for the Lorado operation throughout both the construction and operating lifetime of the mine, mill, and campsite.

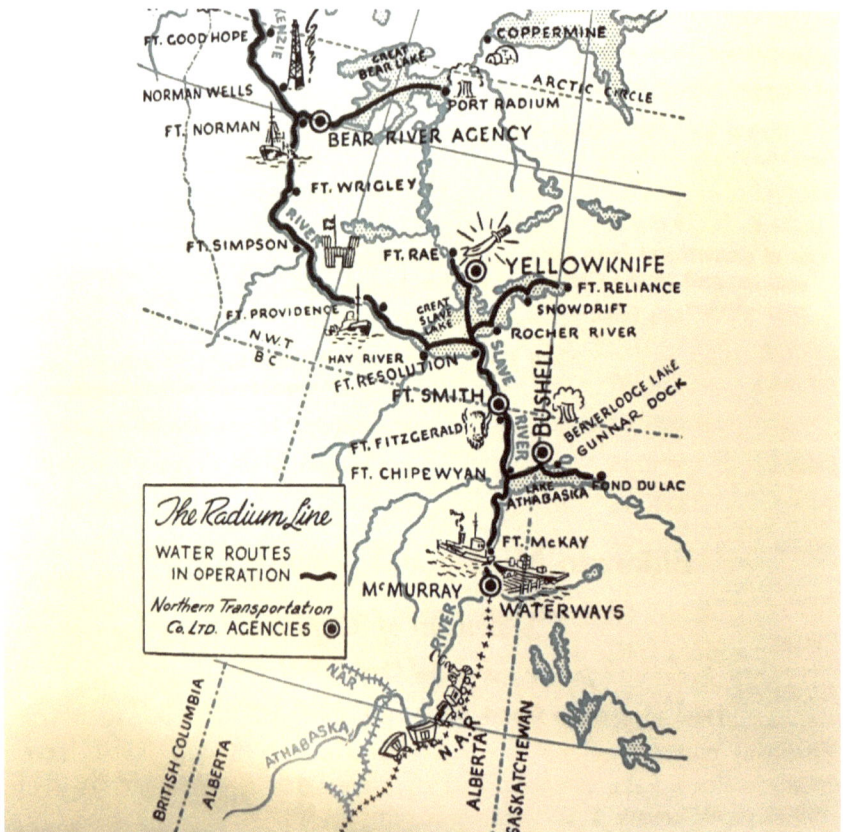

Figure 2.10. Illustration of barge routes northward from the railhead at Waterways, Alberta. Not shown is the leg from Fort Norman to Tuktoyaktuk. Adapted from a drawing in reference [110].

The Radium Line. Northern Transportation Co. operated a series of barge-towing tugboats, called "The Radium Line," that operated along the Mackenzie, Slave, and Athabasca Rivers from Tuktoyaktuk in the north to Waterways in the south [110]. Along the way, barges could traverse Great Bear Lake to the uranium mine at Port Radium, Great Slave Lake to the gold mine at Yellowknife, and Lake Athabasca to the uranium mines in the Uranium City and Beaverlodge areas (Figure 2.10).

Even later, in the 1950s, the cost of freight by air was more than double that by water. As a result, mining companies in this region would attempt to order a year's worth of supplies prior to the beginning of the boat/barge season[26] for delivery in advance of winter freeze-up [1]. As a "rule of thumb," it was found that both capital and operating costs for the Athabasca region mines and mills, including Lorado, tended to be more than double those of comparable southern operations [1].

Figure 2.11. The Northern Transportation Co. boat *Radium Prospector* with a barge, carrying freight on *"The Radium Line"* in 1960. Library and Archives Canada (Photo 4952391).

---

[26] Ice breakup would usually occur in late May, with some areas not clearing until mid-June, and the lakes would usually be frozen-over by the end of October yielding an effective barge season of about 15 weeks each year [15,31].

A consequence of the high cost of transportation by water was Lorado's decision regarding the source of sulphur for its acid plant. Lorado decided to mine and process its own pyrite rather than ship-in elemental sulphur (as Gunnar Mines did) [82]. When Lorado began experiencing shortages of sulphuric acid in 1958 (see below), the company purchased more from Gunnar Mines then had to wait until the ice on Lake Athabasca was thick enough to support its transportation. Even then, additional quantities were shipped in order to have enough stockpiled to see the company through the ice break-up period, after which additional acid was delivered by barge [101].

Figure 2.12. Unloading freight at Bushell, *circa*. 1957. Courtesy of the University of British Columbia, Rare Books and Special Collections (Northern Miner fonds).

2.3 Construction and Initial Operations.

The mine infrastructure had evolved over the years since 1953 (Figure 2.13). The mill construction was launched in June 1956 (Figures 2.14-2.17), while the rock crushing operations were initiated in March 1957 [82,111]. Both mine and mill were considered to be fully operational by May 1957[27], at which time ore shipments from Cayzor Athabaska and Lake Cinch were initiated. These were followed by shipments from National Explorations in June 1957.

Figure 2.13. View of the Lorado mining plant in 1957. Beaverlodge Lake appears in the background (Lorado Uranium Mines Ltd. [111]).

---

[27] The official ceremonies to mark the start of production were held on August 5, 1957.

Figure 2.14. View of the mill site under construction in 1956 (Lorado Uranium Mines Ltd., [82]).

Figure 2.15. Excavating for the jaw crusher. Note the residences in the background (Lorado Uranium Mines Ltd., [82]).

Figure 2.16. View of the mill site under construction in 1956, showing foundations being poured for the acid plant, powerhouse, concentrator, crusher house, and storage buildings (Lorado Uranium Mines Ltd., [82]).

Figure 2.17. Views of the concentrator (upper right) and powerhouse (lower left) under construction in July 1956 (Lorado Uranium Mines Ltd., [82]).

There were start-up problems, of course, particularly with the acid plant. It was discovered that there was insufficient grinding capacity to deal with the pyrite ore. The acid plant was also initially plagued with dust problems that were eventually traced to the graphite in the pyrite ore. This was solved by the addition of a flotation process step, to float and remove the graphite, along with the addition of dust collection equipment [111, 112] (Figure 2.18).

In 1957, a new 23 metre (75 foot) headframe was erected, a new hoist and hoistroom were installed, additional loading and hauling equipment were purchased, and the sizes of the ore bins were increased [111]. Also in 1957, numerous expansions and additions were constructed and/or installed, including additional power, compressed air, heating and water supply equipment, living accommodations, and transportation equipment [111].

Several additional problems appeared in 1958. The uranium ores being shipped to Lorado by the nearby, smaller mining companies, contained twice the concentrations of acid-consuming minerals that had originally been estimated, increasing the acid requirements by about 125 kg of acid per tonne (250 pounds of acid per ton) of ore to be treated. To make matters worse, it turned out that the pyrite milling and burning processes were only able to produce about 32 tonnes (35 tons) of sulphuric acid per day, and it was judged that they would never meet the newly projected demand of over 73 tonnes (80 tons) per day [101]. This plus an explosion in the acid plant in December of 1957 caused acid shortages that were dealt with, in the short term, by purchasing sulphuric acid from the Gunnar Mine.

To deal with the acid demand problem, management decided to convert the pyrite burning plant to elemental sulphur-burning, and increase the sulphuric acid output capacity to 90 tons per day. This required the construction of a sulphur melter, new storage facilities (for the sulphur), a new waste heat boiler, and other equipment such as pumps [101]. As a result, the pyrite circuit was discontinued in August 1958, and the existing stockpile of pyrite was used up [79]. The new equipment and overhaul of the entire mill were completed in 1959 (Figure 2.18), with both acid plant and milling plants reaching the capacity to process 680 tonnes (750 tons) per day of raw uranium ore with about 93% recovery efficiency [113].

Lorado also discovered that its method of tailings disposal was causing an unexpected problem. The tailings were being deposited into a small lake located close to the west end of Nero Lake [79]. This soon overflowed, allowing the tailings to flow into Nero Lake itself. By disposing of the tailings in a manner that caused them to contaminate Nero Lake, the company was also unwittingly contaminating its own water supply for the milling plant.

Figure 2.18. The Lorado mill, with expansion construction underway, in 1957 (Lorado Uranium Mines Ltd. [111]).

The Provincial Government was also aware of this problem. A Nov. 7, 1957 status report[28] assessed the situation as follows:

"... The tailing lake is about 600 feet from Nero Lake, this 600 feet being muskeg lying almost at lake level. It is only about 200 feet from Nero Lake into Beaverlodge [Lake] and this also is low muskeg. Recently the small tailings lake became so full that the tailings have now begun to run across the strip of land and into Nero Lake directly ... The mill tailings ... waste has a pH of 2.5-2.8. Lorado have kept a constant check on the acid content of Nero Lake as they are drawing water from this source and have found that Nero [Lake] is running from pH 5.7-6.3. This lower figure is apparently quite close to the limit at which the water will be unfit for their machinery or for human use ... I am quite sure that they will have to use a different water source no later than next spring ..."

As predicted, Lorado quickly found the contaminated water, with its increased concentrations of acid, chlorides, and sulphates, almost immediately caused problems in the mill, including equipment corrosion, reduced uranium recoveries, and reduced uranium concentrate quality [101]. Rather than change its tailings disposal practices, the company simply found a different source for the water. A new water supply line was built to direct water from Beaverlodge Lake, which went into operation in October 1958 [101,113].

The Provincial Government was also aware of the consequences of this action. The above-noted Nov. 7, 1957 status report went on to conclude:

"I cannot see any alternative tailing site in this location and ... I believe that we should insist that Lorado be responsible for constructing some kind of impervious dyke across the low strip between Nero and Beaverlodge Lakes to assure that Beaverlodge [Lake] is not contaminated."

The "impervious dyke" constructed by Lorado became known as the "land bridge" between the two lakes. As will be discussed in Chapter 8, the need to protect Beaverlodge Lake would come to the fore in later years.

Lorado's ore reserves were stated in 1956 to be approximately 136,000 tonnes (150,000 tons). The mill expansion to 680 tonnes per day (750 tons per day) (Figure 2.18) was completed in November 1957. By 1959, Lorado's ore reserves estimate had been increased to slightly more than 180,000 tonnes (200,000 tons) of uranium ore [17].

---

[28] Correspondence from D.M. Taylor to C.S. Brown, "Lorado Mill Tailings," Department of Natural Resources, November 7, 1957.

2.4 Wind-Up.

On March 26, 1960, Eldorado cancelled the Special Price Contract for the sale of uranium-bearing concentrates, after which all mining and milling operations were halted [114]. The reason Eldorado stopped buying uranium ore was that, in 1959, the markets for uranium collapsed when the United States and United Kingdom stopped stockpiling uranium. Although the Canadian government continued to buy uranium to preserve some of the industrial activity, most of the uranium mines in Canada closed – Lorado Uranium Mines Ltd. included.

At about the same time, and for the same reason, Lorado cancelled all of its milling contracts with other uranium producers. The balance of Lorado's production contract with the Eldorado refinery in Port Hope was assumed by Eldorado's nearby Beaverlodge uranium mine [115].

The closing of Lorado's mine and mill had an immediate cascading effect on the smaller uranium producers that had been shipping their ore to the Lorado mill, particularly Cayzor Athabaska, Lake Cinch, and Rix-Athabasca, all of which were forced to shut-down and close [116]. Others of the small producers that had been shipping ores to Lorado, such as Black Bay, National Explorations, St. Michael, and Uranium Ridges, had already closed and been shut-down in 1958-1959 due to depletion of their ores. This left Eldorado and Gunnar as the only remaining active uranium producers in Saskatchewan at the time (see Table 1.1).

At the time Lorado ceased mining, the company estimated there was still about 184,000 tonnes (202,893 tons) of "probable" uranium ore on the property, that had been outlined at the lower levels of the mine [114].

Following the closure of the Lorado mine and mill, the company turned to other business opportunities, including participating in prospecting operations in the Northwest Territories, Quebec, and Ireland, and a British Columbia mining operation [115,117].

In 1961, Lorado Uranium Mines also incorporated a wholly-owned subsidiary named Lorado of Bahamas Ltd. which acquired an interest in The Grand Bahama Development Co. (Devco). According to some media reports, Lorado's aim was to participate in the developing of a major gambling resort in the Bahamas [118,119].

In subsequent years the exploration efforts extended to other countries, however Lorado's principal business activities, through its Lorado of Bahamas subsidiary, focused on real estate developments in still other countries [120-124].

Lorado Uranium Mines Ltd., along with several other companies, was amalgamated into International Mogul Mines Ltd. in 1968 [125]. Several mergers and acquisitions later, the Lorado mine and mill came to be owned by EnCana Corp., as illustrated in Table 2.1.

Table 2.1. Lorado Uranium Mines' Corporate Evolution[*] **[17,79,115,125,126].**

Year	Corporate Change
1950	Alco group of seven claims staked by prospector Gunnar Berg for Noranda Mines Ltd.
1952	Property sold to K.S. Oliver, who formed Lorado Uranium Mines Ltd. that same year
1960	Lorado ceased mining and milling and closed the mine and mill
1961	Lorado of Bahamas Ltd. formed as a wholly-owned subsidiary of Lorado Uranium Mines Ltd.
1968	Lorado Uranium Mines Ltd. amalgamated into International Mogul Mines Ltd.
1981	International Mogul amalgamated into Conwest Exploration Company Ltd.
1995	Conwest was acquired by Alberta Energy Company Ltd. (AEC).
2002	AEC merged with PanCanadian Energy Corp. to form EnCana Corp.
2007	Encana transfers responsibility for the Lorado site to the Province of Saskatchewan

[*]See also: "Canadian Corporate Reports," McGill Digital Archive, McGill University, Montreal, 2018, https://digital.library.mcgill.ca/hrcorpreports.

3 THE LORADO URANIUM MINE

3.1 Establishment of the Mine.

Probably the first comprehensive geological description of the area spanning the Lorado Mine was given by Alcock in 1936 [14], with slightly more recent descriptions being given by others [8,15,17,80,81,127-136].

The consolidated rocks in the Lake Athabasca region are all of Precambrian age, and those at the Lorado Mine site are mostly of Archean, early Precambrian age (Tazin Group). The Tazin Group rocks (shown in light blue in Figure 2.3) are generally limestone, dolomite, and secondary silicate rocks [14]. The age of the pitchblendes in the Beaverlodge area has been estimated at approximately 750 million years [137].

The Lorado Mine area geology is described by Macdonald and Kermeen [131,132] as having complex folding, comprising a *"complex of graphitic, chloritic, and sericitic schists interbedded with narrow beds of quartzite. These generally strike northeast and dip at 60 deg. to the east."* Macdonald and Kermeen further note that most of the uranium minerals followed a syncline *"plunging flatly to the north at about 25 deg."*

The Lorado orebody extended along the Crackingstone Peninsula [80]. It was bounded on the east side by quartzite, on the west side by a pyritized shear zone, and it was underlaid by quartzites, phyllites, and schists of the Tazin group [80]. The deposit was somewhat unusual in that it contained virtually no hematite or feldspar, minerals that had occurred in significant quantities in the other uranium deposits in the region [133].

The principal uranium mineral was uraninite. Samples of the uraninite grains were found by Robinson [81] to contain 51% U_3O_8, 4.5% ThO_2, and 16.4% PbO.

Figure 3.1 shows a geological map of the Lorado Mine area. Figure 3.2 shows a generalized plan of the orebody on the first level and a corresponding section through the orebody.

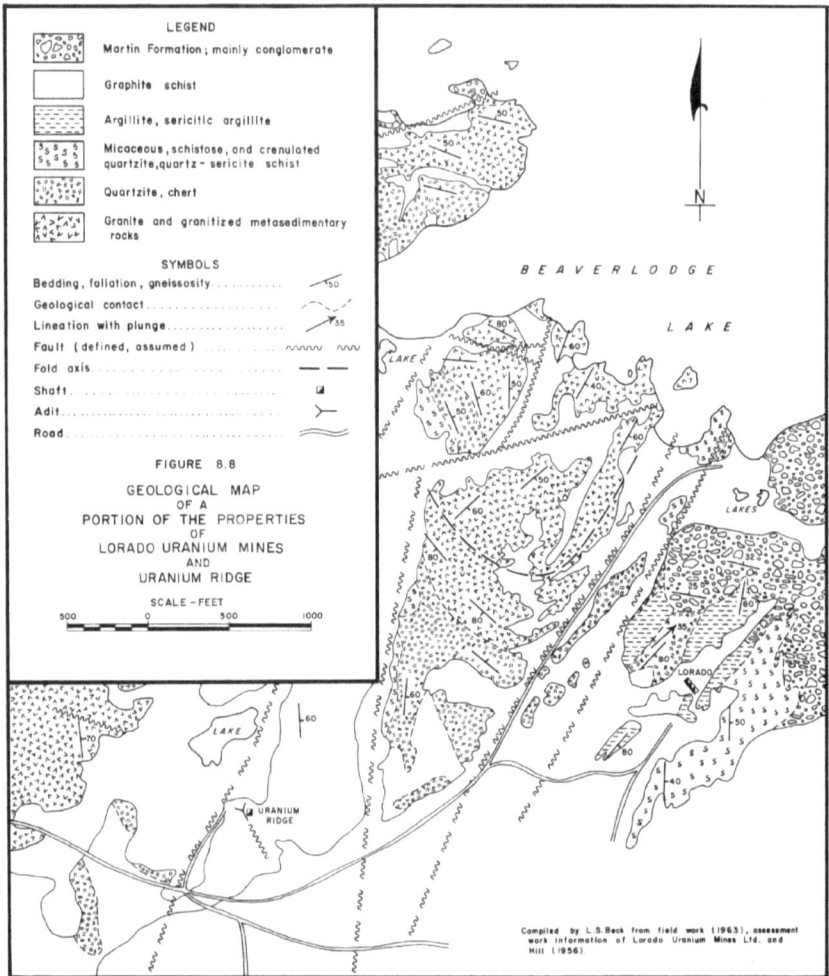

Figure 3.1. Geological map of the Lorado Mine area. The Lorado Mine is shown in the lower right corner. (Saskatchewan Geological Survey, 1969 [17].)

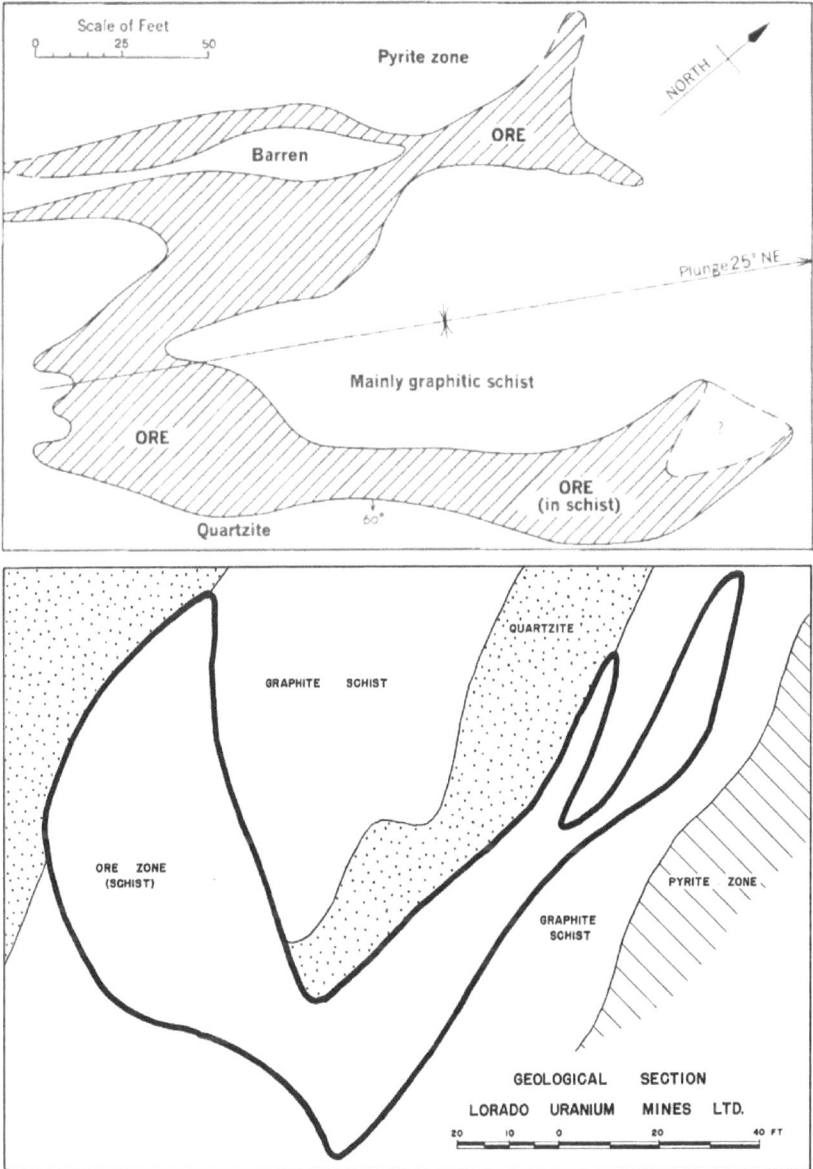

Figure 3.2. Generalized plan of the Lorado orebody on the first level (upper map), and a corresponding section through the orebody (lower map). (Saskatchewan Geological Survey, 1969 [17]; Lorado Uranium Mines, *circa.* 1960; and Geological Survey of Canada, 1962 [8,80])

Perhaps the most unusual feature of the Lorado uranium deposit is that there was also an adjacent pyrite deposit. In 1956, Mamen [103] wrote: *"Lorado is so far unique amongst uranium mines in Canada by having a large sulphide deposit containing from 25 to 40 percent pyrite."* The pyrite deposit occurred just to the west of the uranium ore bodies, in a zone of schists that extended for a depth of at least three levels [15,103,131]. This is shown in Figure 3.2.

It was ultimately determined that the pyrite deposit measured about 180 metres (600 feet) in length, by about 30 metres (100 feet) on the "360 Level." As will be described further below (see Chapter 5), the pyrite deposit was mined to provide starting material for the manufacture of sulphuric acid.

Underground work began in June 1954, with the construction of the main shaft to a depth of 73 metres (240 feet) and lateral development on a single level at 64 metres (210 feet) [103]. In 1955 the main shaft was further developed to a depth of 186 metres (610 feet), and lateral developments (cross-cutting and drifting) were begun at depths of 64 metres (plus sub-levels above and below the "210 Level"), 110 metres ("360 Level"), and 152 metres ("500 Level") [103].

A 23 metre (75 foot) headframe, with collar house, hoistroom, and 136 tonne (150 ton) bins for ore and waste were built at the surface [15,103] (see Figures 2.13, 3.3, and 3.4).

Figure 3.3. The Lorado mine in August 1956. Courtesy of Saskatchewan Archives Board (Photo R-B5998-2).

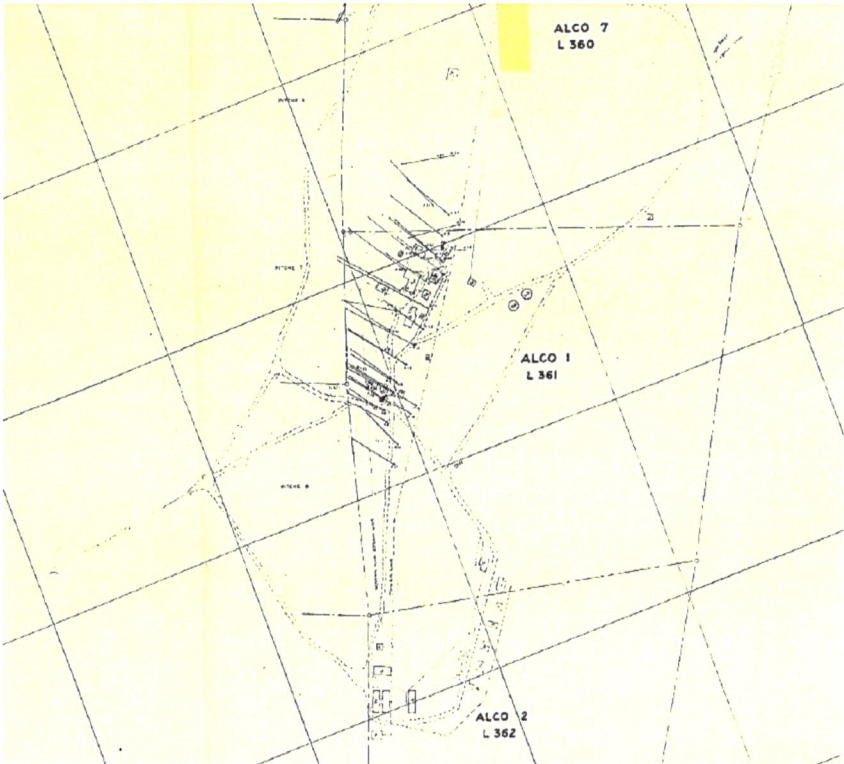

Figure 3.4. Enlarged section of the surface map (Figure 2.4) showing the location of the mine buildings (in the upper left of Alco claim 1) and some of the campsite buildings (in the upper left of Alco claim 2). Lorado Uranium Mines Ltd., 1959.

3.2 Mining[29].

The centrepiece of the surface plant was the 23 metre (75 foot) headframe and hoist [15]. Below this was the main mine shaft. This main mine shaft was comprised of three compartments, two cage/skip compartments, and a personnel access, utility, and maintenance compartment [103]. The larger two compartments (see Figure 3.5) contained cages suspended from a Manitoba Bridge and Iron Works 91 cm by 107 cm (36 inch by 42 inch) double drum mine shaft hoist with a 60 h.p. motor, and 19 mm (3/4 inch) steel wire ropes. The rope speed was approximately 3.8 metres per second (750 ft/min) [15], and the shaft and hoist system had a lifting capacity of about 455 tonnes (500 tons) per day [108].

Two different "cages" were used. A steel personnel cage, capable of carrying ten miners, was balanced by an aluminum skip, having a rated ore-carrying capacity of 1.8 tonnes (2 tons) [15]. The two cage compartments operated like elevators, one transporting miners and equipment up and down, and the other to lift mined ore and waste rock ("muck") up to the surface (see Figure 3.6). The third, "access" compartment was primarily for utility access (air, mine water, electrical, and ventilation) and maintenance, but would also have provided an emergency exit.

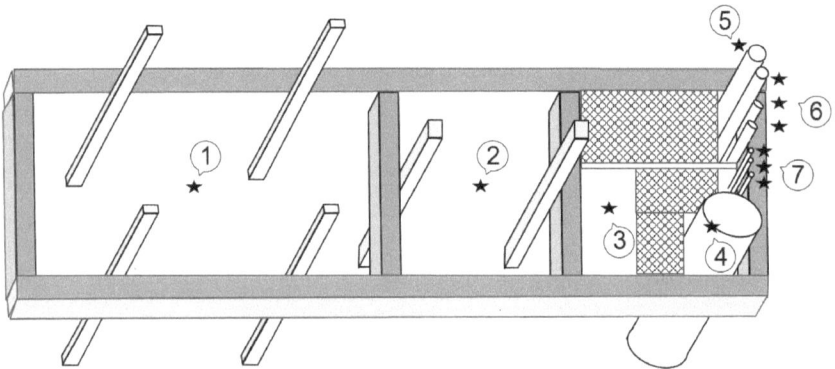

Figure 3.5. Illustration of the probable arrangement of Lorado's underground mine shaft compartments, showing the cage and skip compartments (1 and 2), emergency exit (3), ventilation (4), high-pressure air (5), water supply and drains (6), and electrical conduits (7). Drawn based on specific information from reference [103] and period information from references [70,114].

---

[29] The first Lorado Mine Manager was G.D. Oliver [138].

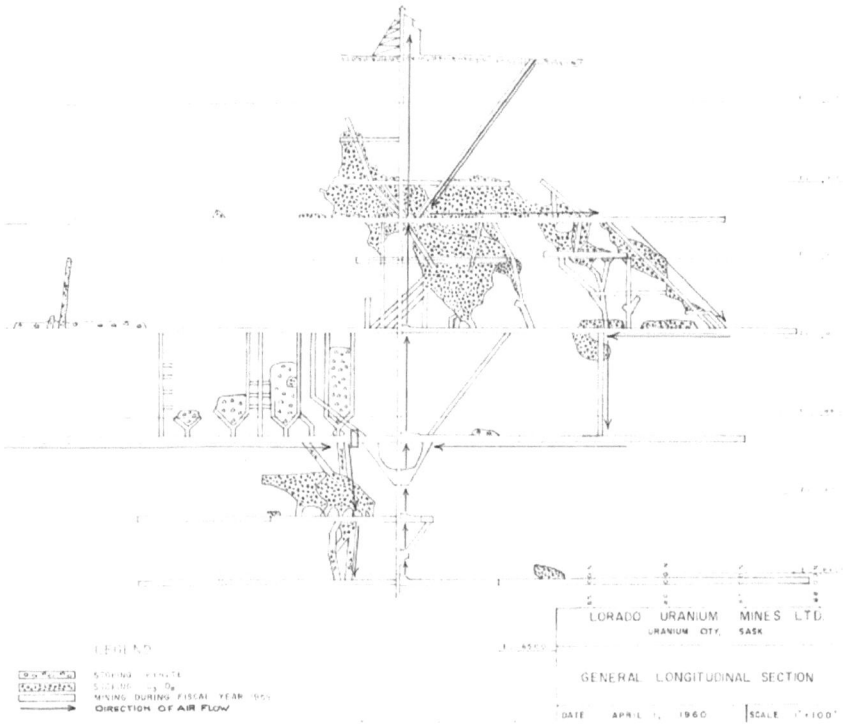

Figure 3.6. Excerpt from a general longitudinal section showing most of the mine workings. The arrows show the air flow directions. Note the four pyrite mining areas shown to the left of centre. (Lorado Uranium Mines Ltd., 1960). The complete drawing is shown in Appendix 1.

Ventilation was provided by having fresh air drawn down through raises and then returning up the shaft. The airflows are shown in Figure 3.6 and Appendix 1. Electric fans (12" Meco fans) were used to ventilate development headings [103].

Two underground mining methods were used at Lorado, "*horizontal cut and fill stoping*" in the larger ore zones, and "*shrinkage stoping*" in the smaller ore zones [15,103]. Stoping is a process wherein ore is removed from spaces or passageways called drifts, which produces open spaces called stopes.

Figure 3.7 shows a stope plan and section, and Figure 3.8 shows a 1960 illustration of where each mining process was used. Transverse stopes were laid out at 12 metre (40 foot) widths with 9 metre (30 foot) pillars between them [15]. Service raises were driven through the pillars from level to level [15].

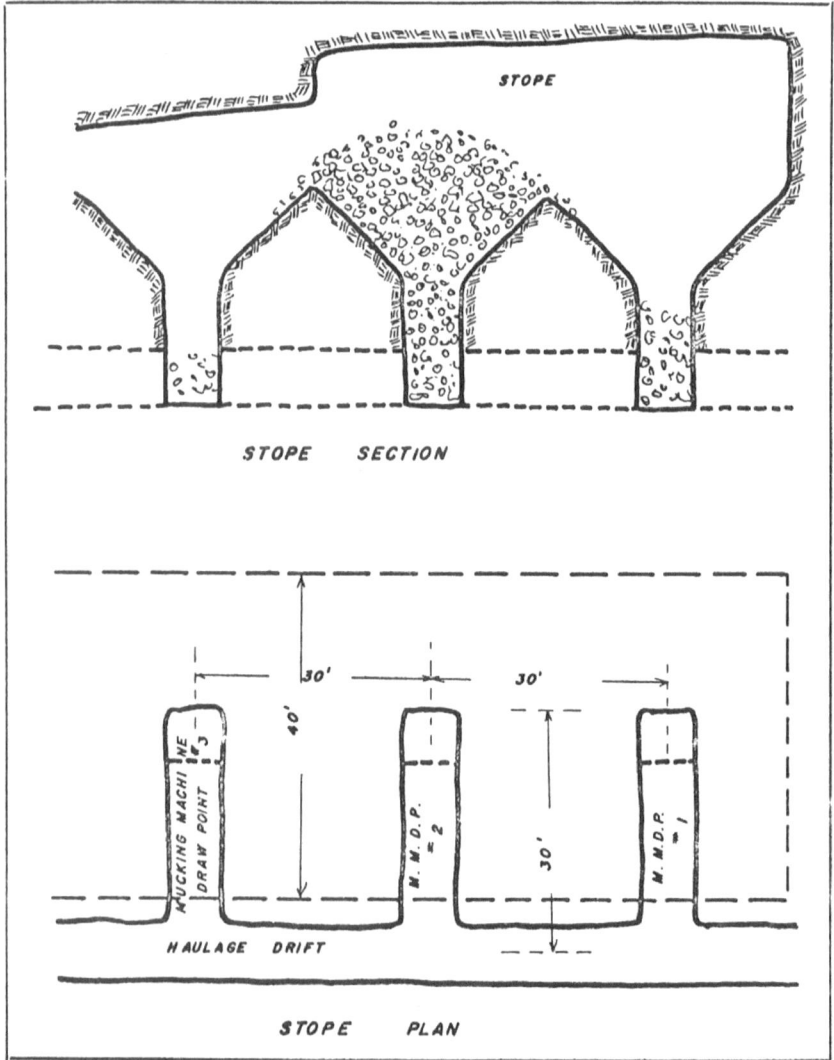

Figure 3.7. Example of a stope section (upper) and plan (lower) in the Lorado mine. (Lorado Uranium Mines, *circa*. 1960; and Geological Survey of Canada, 1967 [8])

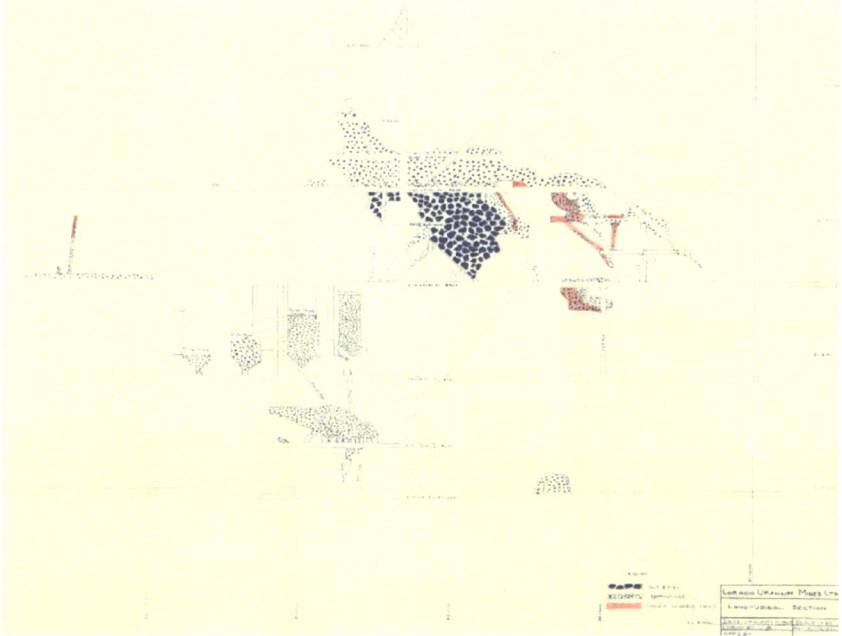

Figure 3.8. Excerpt from a general longitudinal section showing most of the mine workings. This drawing shows where horizontal cut-and-fill stoping was used (on the second level), and where shrinkage stoping was used (on all levels). (Lorado Uranium Mines Ltd., 1960). The complete drawing is shown in Appendix 2.

In stoping, miners would drill a pattern of holes, fill the holes with explosive, and then blast to break-up the ore[30]. The blasted, broken-up ore could then be mined and transported. These methods were possible because the surrounding granite rock was stable enough not to collapse after the ore had been mined out, although the stopes would have been stabilized with timbers bolted to the rock.

Cut and fill stoping involves the mining of horizontal slices of ore, beginning with the bottom slice. Once a slice was completed, the excavated area was backfilled with waste rock and mining commenced on the next higher slice, using the backfilled layer as the new working platform. The advantages of this method include ease of following the shape of the orebody, the ability to selectively mine, and the ability to immediately use waste rock for backfilling.

---

[30] An illustration of the process of underground uranium mining in this area and time period is given using film footage from the nearby Rix Athabasca mine in a TMC documentary [63].

In shrinkage stoping, the deposit was blasted and mined upwards from the bottom, in horizontal layers. The broken-up rock was then allowed to fall and remain in place at the bottom of the stope. In this case, the broken rock provided both a working platform and a means to help stabilize the stope[31] [103].

Lorado's pyrite zone was mined by drilling horizontal rings of holes, blasting, and conducting shrinkage stoping. The sulphide ore was mucked from draw points with mucking machines and delivered to shaft pockets by battery-powered locomotives [15,103].

The uranium orebodies were delineated by diamond drilling involving two-miner crews using Copco airlegs and Canadian Ingersoll-Rand JR 38 stoppers and Coromant steel [103]. It is probable that ore-bearing rock was identified using a down-hole Geiger Counter and drill cuttings were collected and sampled for chemical analysis and ore grading[32]. The underground developments were characterized as being "*irregular masses dipping from 30 degrees to 65 degrees having a width of up to 140 feet and lengths up to several hundred feet.*" [103].

Lateral headings were driven at about 1.8 m by 2.1 m (6 ft by 7 ft), with a crew of two mucking, drilling, and blasting a 2.4 m (8 ft) diameter round per shift – each round having a depth of about 2.3 m (7.7 ft). According to Mamen [103]: "*The standard round is drilled with 28 holes without reaming of cut holes as no advantage was derived from this practice.*" Raises were mostly 1.5 m by 2.1 m (5 ft by 7 ft), with a crew blasting 2.4 m (8 ft) rounds [103].

In the 1950s, a common practice at the uranium mines in the region was to blast with water-resistant dynamite explosives such as Forcite (an "ammonia gelatin," usually in 40 or 50 percent grade) or Driftite (a "semi-gelatin," typically 70 percent grade) [139]. High-strength explosives of this type were found to be the most suitable for dealing with hard rock and wet conditions. The specific explosive used at Lorado was Cilgel (at 50 percent grade[33]), using Thermalite tape fuses (igniter cord) [103]. A detailed description of the explosives, ancillary equipment, and methods used in Canadian underground mines in the late 1950s is given by Dyment [139].

Access-ways and working platforms were probably constructed from planks and ladders, supported by steel bolts drilled into the walls. Ore shoots were likely located by diamond drilling upwards at 7.6 metre (25 foot) intervals and then testing the recovered cores (or sludge)

---

[31] However, some of the blasted rock has to be removed from the drift as mining progresses, because broken-up rock takes up more volume than it did when intact.

[32] An illustration of the processes of diamond drilling ore grading in this area and time period is given in a TMC documentary [63].

[33] 50% nitroglycerin mixed with cellulose, sodium or potassium nitrate, and a hydrocarbon like tar (to make it waterproof).

radiometrically. A typical practice would have been to take muck samples from each skip hoisted to the surface, and to also test them radiometrically. As was common practice in the region, Lorado back-filled mined-out drifts with waste rock to further enhance stability.

According to Mamen [103] a slusher, comprising a double drum hoist and a scraper, was used to drag ore from draw points and pillar stopes. In this method, one cable was attached to the front of the scraper bucket so that it could be pulled along a drift, while a second cable would be anchored at the draw point or pillar used to return the scraper bucket for another load. The Lorado mine had a Canadian Ingersoll-Rand HNNIJ slusher driving a 0.9 m (36 inch) Eimco folding blade scraper [103].

On the main underground levels, ore and waste rock were transported in tram cars, propelled by either battery-powered locomotives (Adas type J) or compressed air-driven "air trammers" (Eimco Model 401, see Figure 3.9) which moved on a narrow-gauge railway having 18-inch gauge rails. The tram cars were 1.4 tonne (1.5 ton) Wabi side-dump cars having approximately 0.4 m³ (13 cu ft) capacity. The tram cars were loaded using rail-mounted compressed-air rocker-shovel mucking machines (Eimco 12B Loader, and Gardner Denver 9), whose front-mounted shovels could load from the front or sides of the track and then throw its contents into an attached tram car. For more details, see reference [103]. The ore and waste rock were transported by rail along the haulage levels to the main shaft for hoisting to the surface (Figures 3.6 and 3.8 and Appendices 1-2).

Freshwater was piped down the mine shaft. The source for this water was Nero Lake, initially, but after discovering that the lake had been seriously contaminated with mill tailings, the company switched to drawing fresh water from Beaverlodge Lake (in 1958).

Controlling the ingress of water from seepages would have been the greatest water problem in the mine. Such seepage was a common problem experienced by all the uranium mines in the region. As mining progressed, drilling and blasting would have constantly exposed ever more fissures and cracks from which water could seep into the mine, threatening to flood the drifts and shaft. The Lorado mine was no exception.

The water problems were dealt with by constant pumping out of the water using a variety of pumps at different levels, including the loading pocket (Fairbanks Morse pump, size 25531 with 10 h.p. motor), first level (Duplex Martin Turner pump), second level (Canadian Ingersoll-Rand Motor Pump, size I-MRVN with 5 h.p. motor) third level (similar Canadian Ingersoll-Rand Motor Pump plus a Model ZGT2 with 50 h.p. motor) [103].

The general mining practice in the mid-1950s was to have automatic level-sensing controls for the pumps, set to maintain the water levels in the drifts at no more than 0.6 to 1.2 m (2-4 ft) from the bottoms [28].

Figure 3.9. Air trammer (left) similar to that used at the Lorado Mine. The one shown is at the nearby Nesbitt-Labine Uranium Mine, circa. 1952. Courtesy of the University of British Columbia, Rare Books and Special Collections (Northern Miner fonds).

The cold was also an issue for all of Canada's northern mines, and heat emanating from the mining machinery, bolstered by underground heaters fueled by Bunker heating oil, combined with circulation aided by fans, was generally utilized to maintain practical working conditions [28].

In Section 1.4 above, it was noted that the health effects of low doses of radiation were not yet well known in the 1950s, so uranium mining was still considered "safe" (as far as radiation hazards were concerned). In the television documentary film, *"The Road to Uranium"* [65], a reporter speaks to one of the underground uranium miners at the nearby Eldorado mine in 1957, and asks *"Is uranium dangerous to work with? Do you have to be careful?"* to which the miner replies *"No … [I] never experienced anything anyway."*

Underground uranium mining was hazardous, however. A constant flow of air had to be provided, not just for breathing, but to dilute the exhaust fumes of diesel-powered mining equipment, blasting gases, along with the dust particles (both radioactive and non-radioactive) created by the mining itself. Although the hazard from breathing radioactive particles was not yet well known, the risk of silicosis from breathing silica particles was understood [28]. Air pressure, fans, and large ventilation pipes were used to ventilate the mine workings, unfortunately not always very effectively. An examination of one of the nearby Eldorado Beaverlodge mines (probably Ace or Fay) in 1954 identified "*active dead ends at the mine with no ventilation*" [28].

Figure 3.10. A geologist testing a uranium ore exposure with a Geiger Counter at the nearby Eldorado Beaverlodge mine in 1960 (reference [110]).

3.3 Mine Evolution.

By the end of 1957, considerable lateral work had been completed at first through third levels, at about 67 m (220 ft), about 110 m (360 ft), and about 150 m (500 ft), plus on several sub-levels, and including pyrite mining on the third level (see Appendices 3-6) [111]. In 1958, work on all of these levels was continued, with the second level being extended to the west and minor extensions to levels one and three.

Also in 1957, the fourth and fifth levels were opened at about 183 m (600 ft) and about 206 m (676 ft) (see Appendices 3, 7, and 8) [101]. Significant extensions to the fourth and fifth levels were made in 1958. Figure 3.11 shows an excerpt from a mine composite plan showing most of the mined drifts, at various working levels.

All of these developments continued in 1959 through 1960, particularly on the fourth and fifth levels (see Figures 3.6 and 3.8 and Appendices 1-2) [113].

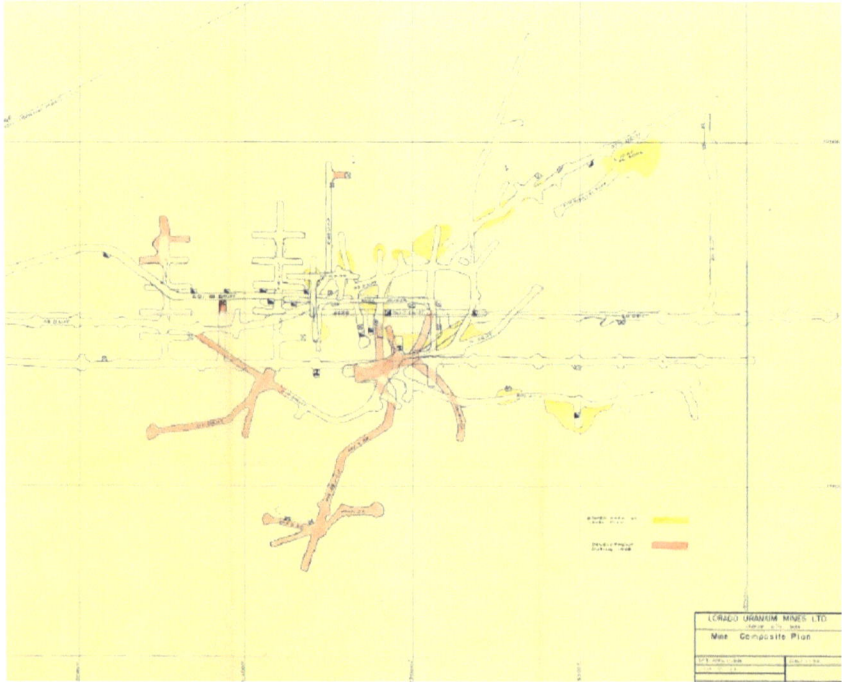

Figure 3.11. Excerpt from a mine composite plan showing most of the mined drifts, at various working levels. (Lorado Uranium Mines Ltd., 1959). The complete drawing is shown in Appendix 9.

3.4 Waste Rock.

Any waste rock not used in stoping or back-filling of mined-out drifts was hoisted to the surface, although the actual amounts hoisted are not known.

Photographs of the mine site during its operating years show at least one waste rock pile situated to the west of the head-frame (as can be seen to the right of the headframe in Figure 3.3).

Several references to the status of the mine after closure assess the amount of residual waste rock at little to zero (e.g., [140]), so it is possible that the bulk of the produced (hoisted) waste rock was used in road- and foundation building, in conjunction with some of the clean-up activities after closure. When the Lorado mine and mill buildings were dismantled in 1982 and 1990, respectively, they were buried using waste rock (see Chapter 7).

When site inspections were conducted during the final remediation phases (see Chapter 8) only small piles of residual waste rock, typically of the order of 1 m^3, were found.

3.5 Mining Retrospective.

As already noted, on March 26, 1960, Eldorado cancelled the Special Price Contract for the sale of uranium-bearing concentrates, after which all mining and milling operations were halted [114]. Over its operating history, the Lorado mine produced a total of about 104 thousand tonnes (115 thousand tons) of uranium ore (not including the uranium ores obtained from other mines) and about 54 thousand tonnes (60 thousand tons) of pyrite for the mill (see Chapter 5). Most reports suggest the average grade of ore produced from the Lorado mine was about 0.2% (as U_3O_8) [86-88,91].

Table 3.1. Lorado's Uranium and Pyrite Mining History [101,111,113,114].

Year Ending	Uranium Ore Mined and Milled (tons)	Uranium Ore Milled from Stockpile (tons)	Total Uranium Ore Mined and Milled* (tons)	Pyrite Ore Mined and Milled (tons)
April 30, 1957	11,696	7,075	18,771	19,684
April 30, 1958	29,810	4,026	33,836	33,023
April 30, 1959	30,368	1,681	32,049	7,895*
April 30, 1960	28,374	1,655	30,029	0*
Totals	100,248	14,437	114,685	60,602

* Does not include uranium ores milled from other mines (see Table 5.2).

4 MINE INFRASTRUCTURE AND THE LORADO CAMPSITE

The Lorado site is somewhat remote, existing in a sub-arctic region that is semiarid with short cool summers and cold winters. Being located near the shore of a large lake in Saskatchewan's north, the site was subject to almost constant winds. Temperatures range from cool in the summer, (averaging 17 °C in July, with a maximum recorded high in the region of 35 °C in 1984), to very cold temperatures in the winter (averaging -27 °C in January, with a maximum recorded low in the region of -49 °C in 1974) [68,141].

These are the officially recorded numbers, however, earlier reports refer to lower temperatures. In his autobiography [142], prospector and developer Ted Kennedy provides a description of life in the nearby Nicholson camp, in January of 1959:` *"It's cold in the winter in that country, and that winter was no exception. It got to be 60 below on a few nights and 50 below was common."*

The uranium companies in the region were able to attract a workforce to remote locations such as this only by paying close to the highest wages in the mining and mineral industry in Canada [143].

In 1952, the provincial government decided to establish Uranium City. This was the same year in which Lorado Uranium Mines was formed, enabling Lorado to establish its official mine and mill offices in the city. Lorado's employees either lived in Uranium City or, once it was constructed, the Lorado camp-site.

Some descriptions of Uranium City and life in the town in this time period are given in references [39,45-48,144] and illustrated in the 1953 TMC [63] and 1957 ITN [65] documentary films.

4.1 Mine Infrastructure.

As noted in Chapter 3, the mine's surface infrastructure included a 23 metre (75 foot) headframe, with collarhouse, hoistroom, and 136 tonne-capacity (150 ton) bins for ore and waste rock (see Figure 2.13).

Mine Operations. The principal mine-site operations buildings included a mine dry[34], two machine shops, carpenter shop, warehouse, storage shed, and a garage (Figure 4.1). Most of these were of wood-frame construction with shiplap, asbestos mine siding, fibreglass insulation, and Donnacona inside sheeting [103]. Separate sheds were provided for a powder magazine, and for blasting cap and fuse storage. As noted in Chapter 2, the company initially used offsite labs for assaying drill cores, but eventually opened its own onsite chemical assay laboratory in the fall of 1954 [107].

Figure 4.1. Aerial view of Lorado mine infrastructure in 1957. (Saskatchewan Government Photo)

Water and Fuel. Most, if not all, bulk petroleum products, including diesel oil and gasoline, would have been shipped during the open-water season each year. They would have been taken by barge to Uranium City or

---

[34] A mine-site building in which workers can change into and out of their working clothes and wearable equipment. Such facilities generally also include sinks, showers, toilets, lockers, and dirty-clothes baskets.

Bushell and then trucked from there to the Lorado site. Diesel oil was stored in two approximately 450 kL tanks (Black, Sivalls & Bryson Ltd., 100,000 gal each) [103].

Water for the boilers and campsite use was initially obtained underground from diamond drill holes and stored in two 68 thousand litre (18,000 gallon) clean water tanks housed in a pumphouse [103,145]. A chlorination system seems to have been installed in the summer of 1956 [103].

As noted above, the company began to draw fresh water from Nero Lake in about 1956, but after discovering that they had seriously contaminated this lake with their mill tailings, the company switched to drawing fresh water from Beaverlodge Lake in 1958. The water was pumped from the lake through a 25 cm diameter (10") wood stave pipeline.

Power and Heat. All electric power was diesel generated and supplied from the powerhouse, which contained a 6VEB English Ruston 250 kVA generator. Two smaller systems were provided for emergency power needs: a 120 h.p. stationary Fairbanks-Morse engine driving an English Electric 94 kVA generator, and an 87 h.p. Caterpillar engine driving an Electric Repair and Machine Co. 65 kVA generator [15,103]. The generators were housed in a powerhouse (which was a Soulé building) [103].

Plant heating was provided by 50 h.p. and 25 h.p. Napanee oil-fired boilers in a boiler house at the mine, and a similar 25 h.p. unit for the campsite buildings [103].

Compressed Air. Several mining machines, including the air trammer, operated on compressed air. For this, compressed air was provided by a 1000 c.f.m. Copco AR5 compressor driven by a 200 h.p. G.M.C. diesel motor, plus a Canadian Ingersoll-Rand Gyroflo 600 c.f.m. compressor and two Canadian Ingersoll-Rand Rotary Mobil-Air compressors delivering 500 c.f.m. and 210 c.f.m., respectively [103].

Rolling stock for the site, as of 1956, included a D7 Caterpillar bulldozer, TD 14 International bulldozer, Ford 3-ton dump truck, two G.M.C. 1-ton dump trucks, two Chevrolet ½ -ton Pickups, and a Willys Jeep Station Wagon [103].

Other. Although the company originally planned to operate its own boats for freighting materials to and from Eldorado and Bushell, it seems to have instead relied on boats from the nearby Pitch-Ore Uranium Mine [106].

4.2 Campsite Infrastructure.

The Lorado campsite was constructed in two sections, one close to the mine (Figures 4.2 and 4.3) and the other one about 6.4 km (4 miles) away, close to the mill (Figure 4.4). The company began ordering lumber and equipment to begin constructing a "permanent" campsite near the mine in August 1953 [103,104]. In about 1956, construction began on a second section of the campsite, which was located near the mill site. By 1957 Lorado had a staff of 40 working in the underground mine plus 90 in the mill, with its contractor McNamara Construction Co. providing another 100 to work on the mill expansion project [112].

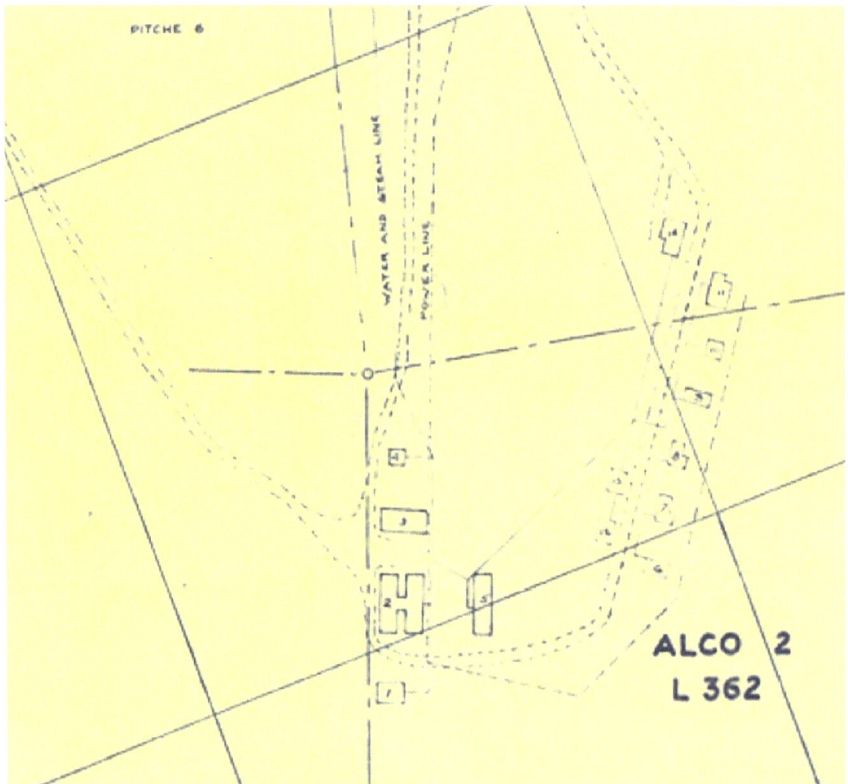

Figure 4.2. Section of a surface plan layout showing the layout of the campsite near the mine. From bottom to top along the vertical-centerline are the coffee shop, bunkhouse, cookery, and boiler-house. To the right of these is shown the staff house, and further to the right are shown a double row of staff residences. Lorado Uranium Mines Ltd., 1959.

The portion of the campsite that was located near the mine consisted of a coffee shop, a large two-story cookery with apartments upstairs, a dedicated boiler-house, 10-person staff house, eleven staff/family residences, a cottage, and a large 80-person, H-shaped, one-floor bunkhouse [103]. Co-located was the main office (separate from the one in Uranium City), and a storage shed. These are all shown in the surface plan layout of Figure 4.2, and the photograph in Figure 4.3.

Figure 4.3. Aerial photograph, *circa* 1957, of the campsite near the mine (foreground). The headframe and mine operations buildings are in the upper left. Courtesy of the University of British Columbia, Rare Books and Special Collections (Saskatchewan Government Photo, Northern Miner fonds).

The portion of the campsite that was located near the mill consisted of at least six more residences, some of which can be seen in Figure 2.15. The general layout for this portion of the campsite is shown in Figure 4.4.

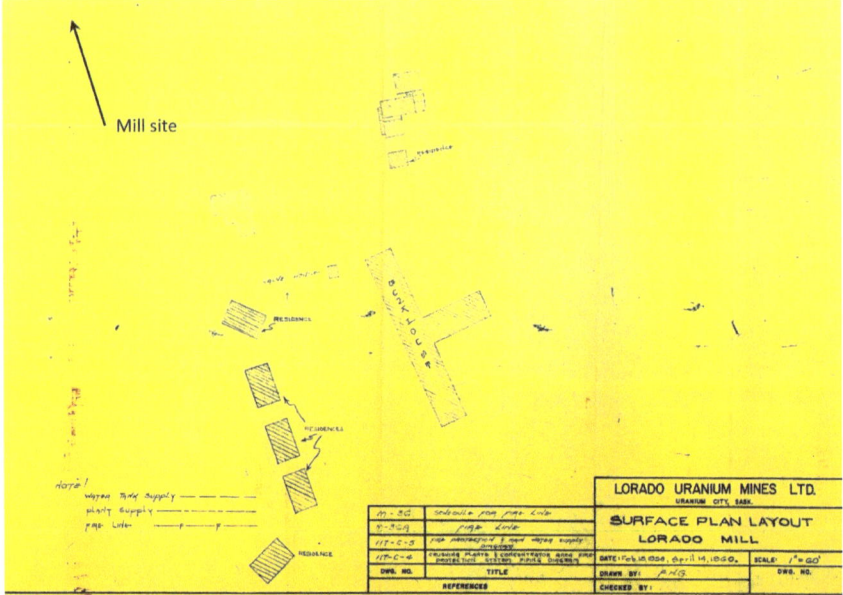

Figure 4.4. Section of a surface plan layout showing the layout of the campsite near the mill. The arrow in the upper left indicates the direction of the mill. Lorado Uranium Mines Ltd., 1960.

The recreational activities in this area, and era, generally included boating, hunting (caribou, deer, and moose), and fishing (trout, jackfish, and grayling) [68, 142, 146, 147].

5 THE LORADO CUSTOM MILL

5.1 Uranium Milling Background.

The Atomic Age and Cold War Era uranium mills were somewhat unique in Canada in that they were in very remote locations (making the shipping of process equipment and chemicals to the mills unusually expensive). With harsh winter climates (creating special problems of heating, moisture, and sometimes permafrost), the mill was required to efficiently process low-grade ores (for which an alternative to the classical "gravity method" of separation had to be developed and then adapted to the particular ore from a given mine). Furthermore, "EMF control[35]" was required to maintain a particular oxidation state on the part of the uranium being leached. The Canadian uranium ores contained pitchblende and uranophane with a large percentage of uranium (IV) [1], requiring the mills to control the oxidizing nature of the leach solutions.

Most of these issues were dealt with in the development of the first two large uranium mills in Canada, namely those at Port Radium and Eldorado-Beaverlodge [1]. The Canadian Department of Mines and Technical Surveys (part of what is now Natural Resources Canada) developed an alkaline leach process that was introduced at Port Radium in 1952, and then adapted for and introduced at Eldorado-Beaverlodge in 1953 [1]. These were the first two commercial-scale, continuous plants for leaching uranium ores in the Western Hemisphere (and the second and third in the world after one in Portugal) [1].

The Eldorado uranium mill at Beaverlodge was built near the Ace-Fay

---

[35] In electrochemistry, EMF (electromotive force) refers to the difference in electrode potential between two electrodes in a cell and is related to the potential for oxidation or reduction reactions to occur. In the present context, EMF control refers to maintaining suitably oxidizing conditions in the uranium leaching circuit described below.

mines and was originally designed to handle 454 tonnes per day (500 ton/day) [10]. It was expanded to 680 tonnes per day (750 ton/day) in 1955, and then again to about 1,800 tonne per day (2,000 ton/day) in 1957 [148]. In addition to processing ores from the nearby Eldorado mines, this mill also processed ores from other, smaller mines in the area, such as Consolidated Nicholson Mines [37].

The reason Eldorado chose an alkaline leach process, was the high carbonate content of ores such as those that were going to be produced from the Ace Mine. In their case, high carbonate-content ore would cause high acid consumption in an acid leach process, so it seemed that an alkaline process might be better, but the company was not confident about this until pilot testing was completed at the Sherritt Gordon facility in Ottawa [10]. Ultimately, the alkaline leach process was adopted for the Eldorado mill, which commenced operations in 1953.

Conversely, more conventional (at the time) acid leach processes were developed for the nearby Gunnar and Lorado mills [15,68,70,143,149,150]). This provided a degree of processing flexibility for the smaller mines in the area, which could contract to send their ores to the Eldorado mill for alkaline leach processing, or to the Lorado mill for acid leach processing. Comparisons of the Canadian acid and alkaline leach processes are given in references [150,151]. By 1954, both kinds of processes had been developed to enable *"profitable treatment of most ores containing from 0.25 per cent to less than 0.1 per cent U_3O_8"* [150].

As already noted, Lorado's mill was located 6.4 km (4 mi) away from its mine, in order to have it reside in a more central location with respect to the neighbouring mines [110]. Its location can be described as being 10 km west (by road) from Uranium City and adjacent to the west shore of Nero Lake, a small lake that drains into Beaverlodge Lake, and ultimately into Lake Athabasca. Figure 5.1 provides an illustration of just how close the mill was to Nero Lake.

Figure 5.2 shows a close-up of the powerhouse – concentrator complex. In this figure, the building in the foreground is the powerhouse. Behind it, and visible to its left and right, is the concentrator building, which contained the leaching, filtration, and ion exchange circuits.

Figure 5.3 shows the layout of the major mill site structures (excepting those of the campsite near the mill, which are shown in Figure 4.12).

Figure 5.1. The Lorado uranium mill site in 1957. Courtesy of Saskatchewan Archives Board (Photo R-B6767-2).

Figure 5.2. The Lorado uranium mill site in 1957. The building in the centre-foreground is the powerhouse. Behind it, and visible to its left and right, is the concentrator building, which contained the leaching, filtration, and ion exchange circuits. Courtesy of Saskatchewan Archives Board (Photo R-B6767-1).

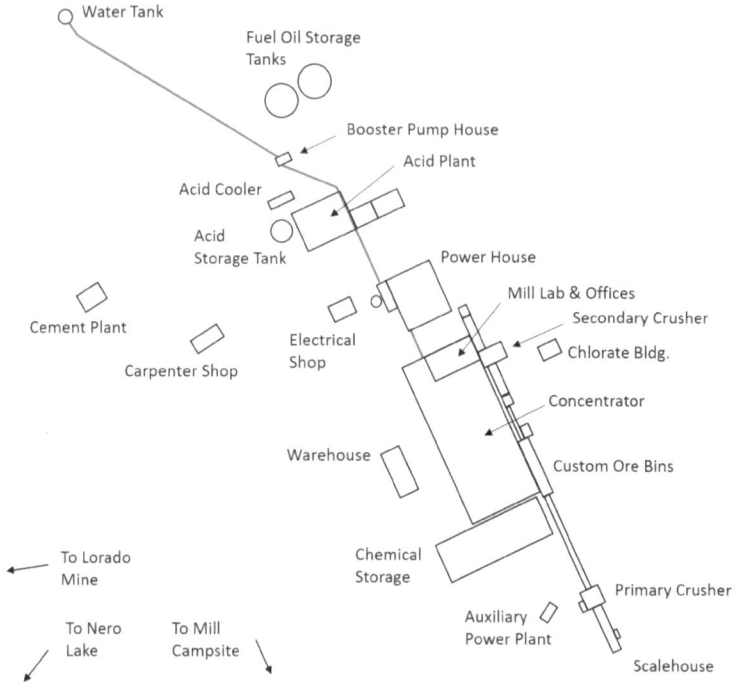

Figure 5.3. Surface Plan Layout for the Lorado Mill (drawn approximately to scale, based on a drawing by Lorado Uranium Mines Ltd., April 14, 1960).

Figure 5.4. The Lorado mill site in 1957. This photo shows the main structures illustrated in Figure 5.3, from the fuel oil storage tanks (to the left) to the primary crusher and scale house (to the right). Courtesy of Saskatchewan Archives Board (Photo S-B3748).

5.2 Uanium Milling at Lorado.

The Lorado mill was designed for flexibility and expandability. The base design was for 454 tonnes of ore per day (500 ton/day) and included provision for storing, sampling, and blending ores from other mines beyond Lorado's.

The reason for expandability was the company's hope that the mill would be able to take *"all the potential small mine production in the district,"* which it was estimated could amount to as much as 900 to 1,360 tonnes per day (1,000 to 1,500 ton/day) [145]. In addition to being necessary for Lorado's mill to be economic, it was expected that having a custom mill in the region would, in turn, enable the development of such smaller mines, which would otherwise not be economic even as mines. As Lorado's Mill Manager Jack R. Woodward noted in 1956 [145]:

"… it is hoped that its operation will give encouragement to the 'small' operators in the industry and contribute to the development of Canada's Mineral Resources."

The mill's completion and opening in the spring of 1957, at its original design capacity, were greeted with some enthusiasm in the media (Figure 5.5) [112,156-158]. Within two months of the mill's opening, it was processing uranium ores from other mines beyond Lorado's own, beginning with ore from Lake Cinch Mines [158]. With other companies lining-up to ship their ores for processing as well, Lorado quickly revised its design and the mill was expanded in phases, completed between 1957 and 1959, to ultimately reach 680 tonnes per day (750 ton/day) [8,113,148,152,155].

According to Woodward, the ore was purchased from the smaller mines based on the assayed uranium contents, although at the time the specific amounts of ore contracted and the prices paid were all classified as "Secret" [145]. Some estimates of the amounts of ore shipped to Lorado from other mines and prospects are given in Table 5.1 [116,130]. Although even the high end of the totals reportedly shipped to Lorado does not match the total Lorado reported actually receiving (see Table 5.2 at the end of this chapter), this does give some idea of the amounts sent by the various other operations in the region.

The Lorado milling process was a "standard" acid leach process and consisted of a number of operations that have been described in some detail in the public literature [1,8,15,67,68,145,148-151,153-155]. Mill operations were conducted with a staff of approximately 96 people working nominally 44-hour work weeks on three shifts, enabling continuous 24 hours per day, 7 days per week operations [10].

Lorado's Mill Starts to Run

The crushing and grinding section of the custom uranium mill of Lorado Uranium Mines has started operation, less than a year from the start of construction.

The custom mill is located in the Beaverlodge area, Northern Saskatchewan. First units of the operation started Mar. 17. Production of sulphuric acid, primary reagent of the milling process, is expected to commence during the first half of this month. Then the concentrator proper will start to process uranium ore and when the mill circuits finally are filled with ore some weeks later, first uranium concentrates will be produced.

Construction of Lorado's mill has been quick. Barges brought equipment to the remote property after breakup last June and in less than nine months from the time first concrete was poured, the crushing and grinding section was at work.

High Quality

But despite the speed of construction, the company believes that a high quality mill has been created, directors state in a current report to shareholders. Construction was done by McNamara Construction Co. and engineering for the project was supplied by Kilborn Engineering and its associates, McCune Engineering.

The mill will have an initial rated capacity of 500 tons per day, but it is expected that expansion to 750 tons per day will be completed by next October. Equipment for expansion already has been ordered for several months.

Commence Leaching At Lorado Uranium

The brand new uranium treatment plant of Lorado Uranium Mines in the Beaverlodge area, Northern Saskatchewan, is gradually being broken in. The company has started up one circuit, handling sulphides for the production of sulphuric acid to be used in the uranium treatment process.

Crushers, ball and rod mills have been running on rock for several days and first ore was fed to the leaching circuit at the end of last week.

Lorado's mill, built in record time, will operate on a customs basis with ore supplied from its own mine as well as those of Cayzor Athabaska Mines, St. Michael's Uranium Mines, Lake Cinch Mines, National Exploration and Black Bay Uranium.

It is expected that Lorado and Cayzor will supply most of the feed requirements to start. Lake Cinch Mines is expected to start shipping some 50-75 tons daily in the near future. The tonnage is made possible through delay in starting shipments from St. Michael's Uranium, which is now expected to start shipping in the fall. Lake Cinch's tonnage may be expected to be increased in the fall when the Lorado mill extension is completed.

Lorado's treatment plant and scheduling of tonnage calls for initial operating rate of about 500 tons per day. This will be increased to a rated capacity of 750 tons per day by October when expansion is expected to be complete. The latter rate is considered adequate to permit the company to fulfill its contract with Eldorado Mining and Refining calling for the production of uranium concentrates valued at $64,-480,000.

Lake Cinch Mines To Start Shipping In June to Lorado

Lake Cinch Mines, one of the few uranium mines in Canada which will go into production free of bond debts, plans to start shipping ore to the Lorado Uranium Mines custom mill in June of this year, President V. R. MacMillan states in the annual report.

The company's mine in the Beaverlodge area of Northern Saskatchewan is about two miles by road from the Lorado mill. Shipments of ore, which will be trucked, are expected to start at 75 tons per day and gradually work up to 150 tons daily. Under its contract with Lorado, Lake Cinch will ship ore containing about 1,500,000 lbs. of uranium oxide (U_3O_8) with a gross value of $15,750,000, the contract to be completed by Feb. 28, 1962.

Ore reserves at the property are computed at about 200,000 tons grading 0.334% (6.48 lbs.) uranium oxide per ton after allowing for mine dilution. The full extent of the orebodies is still not known and several radioactive zones not included in reserve computations will be developed as soon as possible, Mrs MacMillan points out.

Development Work

Operations at the mine, which were suspended for seven months, resumed on Feb. 1, 1957. A 3-compartment shaft was previously sunk to a depth of 648 ft. and 5,681 ft. of lateral work has been done on two levels, the 300 and 500-ft. horizons. About 856 ft. of raising has also been done.

Figure 5.5. Examples of 1957 newspaper articles on the mill opening at Lorado [156-158].

Table 5.1. Estimates of Uranium Ore Shipped to the Lorado Mill from Other Mines and Prospects.

	Uranium Ore Shipped to Lorado		Reference
	(tonnes)	(tons)	
National Explorations Mine – Pat Claims	24,275	26,759	[116]
Eagle-Ace Mine	254-93,810	280-103,408	[116,130]
Strike Uranium Mine	54	60	[116]
Black Bay Uranium Mine	1,247-1,549	1,375-1,708	[116,130]
Beta Gamma Mine	181	200	[116]
Cayzor Athabaska Mine	82,001	90,391	[116]
Lake Cinch Mine	126,285	139,205	[116]
Rix Mine	>513	>566	[116]
St. Michael Mine	>227	>250	[116]
ABC Prospect	64	70	[116,130]
Beaver Lodge Uranium Prospect	68	75	[116]
Strike Prospect	54	60	[116]
Totals	235,224-329,083	259,291-362,752	

Test work on the ores, including some of the process evaluation and optimization, was done for Lorado by the federal government's Department of Mines in Ottawa [145,149]. Work also had to be done to adapt the conventional ion exchange process for recovering uranium from leach solution to produce a higher grade of uranium oxide (yellowcake) suitable for eventual processing into nuclear fuel. Additional process optimization, pilot plant testing, and commercial process troubleshooting were provided by the Saskatchewan Research Council (SRC), beginning in about 1958 [72,74].

Even the conventional methods for uranium assay were not completely suitable. C. Lapointe, of the Department of Mines, had developed a radiometric method for the analysis of uranium that was still in use for mine samples through to 1964, however the uranium mining and milling companies operating in the Athabasca Basin in the 1950s needed a method that was faster, so it could be used for process control in the mills [72]. Dr. Gene Smithson and others (at SRC) developed a new X-ray fluorescence-based analytical method for the determination of uranium in process samples that was both faster and more accurate than the earlier methods, and it was adopted in the Lorado mill [72,74].

A simplified process diagram for the mill is given in Figure 5.6 [145, 155]. The following sections describe the main processing units of the mill, as shown in Figure 5.6. The overall extraction efficiency averaged about 93% (as U_3O_8) of the uranium in the ore introduced into the plant [111,112,150].

Weighing and Crushing Circuit. All of the uranium ores, including those from the Lorado mine, were delivered by truck and weighed on truck scales. The Beaverlodge area uranium ores were considered to be generally medium-hard to hard [153] and were crushed with a jaw crusher and then kept separated until they had been sampled and assayed, after which the material was blended and placed in storage bins. Ore from the storage bins was crushed a second time and then passed through a vibrating screen having 1.6 cm (5/8") openings [155]. Samples of screened ore were collected from the conveyor and sent to the on-site laboratory for analyses [8].

Meanwhile, the blended ore was further crushed and ground in water using rod and ball mills, then finished with spiral classifiers (Figure 5.7). The slurry of ground particles was then thickened in two 15 metre-diameter (50 ft dia.) settling tanks, with Separan[36] used as a coagulant, to create a 55 mass-percent slurry of fine particles (60% less than 0.074 mm diameter, or -200 Tyler Mesh Size) for uranium leaching [150,155].

---

[36] Separan is a high molar-mass polyacrylamide polymer made by Dow Chemical Co.

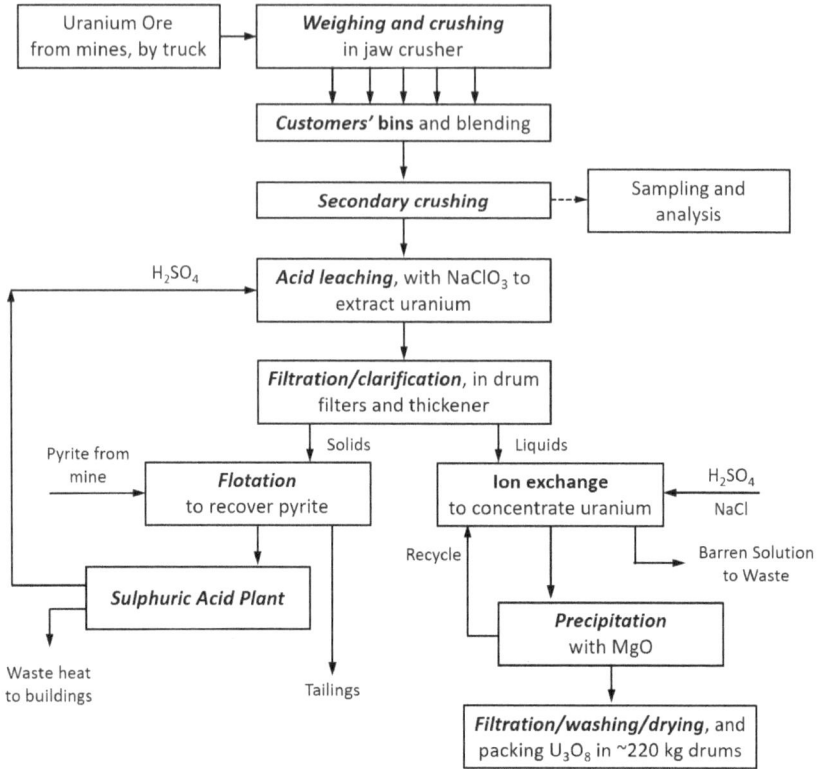

Figure 5.6. Simplified process diagram for uranium ore milling at Lorado's custom mill (Adapted from Woodward [145,155]).

Leaching Circuit. The ore slurry was passed through a series of eight 5.5 by 6 m diameter (18 by 20 ft dia.) agitated leaching tanks. In each tank, the ore was treated with sulphuric acid. The pH was maintained at about 1.7 in order to dissolve the uranium. The agitators were of the "airlift" type, providing direct compressed-air injection into the slurry [155]. Sodium chlorate was also added to the leaching tanks, as needed, to maintain a slight excess in solution, ensuring slightly oxidizing conditions at all times. In this way uranium (IV) was oxidized to uranium (VI), making it soluble in water. The residence time in the leaching tanks was 24 to 36 hours [153]. As noted above, the extraction efficiency was about 93% [111,112,150].

The overall chemical reactions for uranium (IV) and uranium (VI) oxides in the pitchblende ore were:

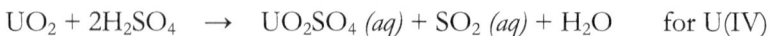

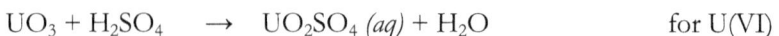

$$UO_2 + 2H_2SO_4 \quad \rightarrow \quad UO_2SO_4 \text{ (aq)} + SO_2 \text{ (aq)} + H_2O \qquad \text{for U(IV)}$$

$$UO_3 + H_2SO_4 \quad \rightarrow \quad UO_2SO_4 \text{ (aq)} + H_2O \qquad \text{for U(VI)}$$

Figure 5.7. Spiral classifiers finish-grinding uranium ore in the Lorado mill in 1957. Courtesy of Saskatchewan Archives Board (Photo R-B7766).

Filtration Circuit. Two stages of filtration and then "clarification" (sedimentation) were used to separate undissolved solids from the dissolved uranyl sulphate. The filtration was accomplished using string-discharge, acid-resistant drum filters[37], at which point a filtration aid was added (a dilute solution of jaguar gum[38]).

The primary and secondary filtrates were combined and pumped onwards for clarification, by sedimentation, in a Dorr Deep Well Thickener tank. The overflow (supernatant) was then polished by further filtration in a Whitco leaf clarifier. The final filtrate ("pregnant solution") was pumped to a 5.5 m diameter by 6 m high (18 ft by 20 ft) storage tank. At this point, the uranyl sulphate was dissolved as $UO_2(SO_4)_3^{-4}$ *(aq)* ions, and still in a solution of about pH 1.8. From the storage tank, the pregnant solution was pumped to the ion exchange circuit.

---

[37] These are drum filters for which a series of strings are drawn across the drum face to dislodge the filter cake.

[38] Jaguar gum is better known as guar gum, a cationic-polymer.

The filter cakes on the drums were washed with 0.25% sulphuric acid solution, re-slurried, and flowed to the flotation circuit (see below).

Ion Exchange Circuit. An ion exchange process was used to recover a high concentration of uranium from the clarified acid-leach liquor (pregnant solution). This circuit employed four I.X. Canada Ltd. ion exchange columns, that would have contained anion exchange resin[39] having chloride ions adsorbed at their exchange sites. The process stream would have been split into two streams, each of which would have been passed "down-flow" (from top to bottom) through each of two of the ion exchange columns in turn. As the solution passed over the resin, the dissolved uranyl sulphate ions were exchanged for the chloride ions. The ion exchange reaction was:

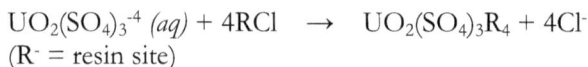

$$UO_2(SO_4)_3{}^{-4} \ (aq) + 4RCl \quad \rightarrow \quad UO_2(SO_4)_3R_4 + 4Cl^-$$
$$(R^- = \text{resin site})$$

Most of the dissolved impurity ions passed through the columns, along with the released chloride ions, and were discharged to the tailings.

The uranium ions were next recovered from the ion exchange columns by flushing them with a 1M solution of sodium chloride in dilute (0.15N) sulphuric acid, which would also restore the columns to their original chloride form. The ion exchange recovery reaction was:

$$4Cl^- \ (aq) + UO_2(SO_4)_3R_4 \quad \rightarrow \quad UO_2(SO_4)_3{}^{-4} \ (aq) + 4RCl$$
$$(\text{NaCl to elute})$$

The effluents, containing concentrated uranyl sulphate, were sent to the precipitation circuit.

Precipitation Circuit. This circuit had three 5 m diameter by 5 m high (16 ft by 16 ft) Denver Agitators. These were large stirred tanks in which the concentrated, dissolved uranium ions were precipitated with a magnesia slurry to form yellowcake[40]. The precipitation reaction was [15]:

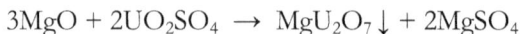

$$3MgO + 2UO_2SO_4 \rightarrow MgU_2O_7 \downarrow + 2MgSO_4$$

This process would have been monitored by pH and would conventionally have been considered to be complete when the pH had reached 7.0. At near-neutral pH, and with oxidizing conditions no longer

---

[39] Most likely a quaternary ammonium "strong-base" type anion exchange resin, of the kind that had been developed in the late 1940s.

[40] Magnesium diuranate has a characteristic bright yellow colour.

being maintained, some uranium (VI) would have become reduced to uranium (IV), yielding UO_2 precipitate as well.

Filtration, Washing, and Drying Circuit. The uranium precipitate was filtered using Whitco and Sweetland (No. 12) plate-and-frame type filter presses that enabled batches of the yellowcake precipitate to be simultaneously washed and filtered. The washing solution was treated with sodium chloride and sulphuric acid to create the solution used for column flushing in the ion exchange circuit above. The filter cake was removed and dried using a Holo-Flite® hollow-screw dryer, in which the slurry was heated as it came into contact with the surfaces of the hollow flights, shaft and trough. While being heated the slurry would have been continuously conveyed, in an axial direction, by the rotating screw flights.

The solid uranium oxide concentrate is commonly known as "yellowcake." It has the approximate chemical formula U_3O_8, but is actually a mixture of uranium (IV) and uranium (VI) oxides and is usually approximately $2/5$ UO_2 and $3/5$ UO_3. The final uranium precipitate was packed into 113 litre (25 imperial gallon) steel drums weighing about 204-227 kg (450-500 lb) each (see Figure 5.8).

Samples of Lorado uranium concentrate that were independently analyzed by the National Lead Company of Ohio in 1958 were found to contain 49 mass% uranium (i.e., 58 mass% as U_3O_8) [159]. The milling process thus increased the uranium concentration by a factor of about 300 (from about 0.2 mass % in the ore to about 60 mass % in the yellowcake).

Shipping. Lorado's drums of uranium (yellowcake) precipitate were trucked to Eldorado, which had constructed an airstrip[41] near its own mines and mill at Beaverlodge in 1951 [68]. This enabled Eldorado to fly the yellowcake directly from its airstrip to Edmonton by an Eldorado Aviation aircraft (Figure 5.9). From Edmonton, the drums were shipped by rail to the Eldorado refinery at Port Hope, Ontario [1,10,109].

---

[41] The Eldorado airstrip eventually became the Uranium City Airport, which was taken over by Transport Canada in the 1970s.

Figure 5.8. Loading a barrel of yellowcake at the Lorado mill in 1957. Courtesy of Saskatchewan Archives Bard (Photo R-A13472-3).

Figure 5.9. Loading yellowcake into an Eldorado Aviation aircraft, for shipment to Port Hope via Edmonton. Eldorado Mining and Refining Ltd., 1956, reference [160].

5.3 Mill Infrastructure.

Dedicated facilities were built for preparing the process chemicals, including mixing areas. These included dissolution and storage tanks, the latter being connected to pumps and piping to deliver the prepared chemical solutions to the appropriate circuits. A complete on-site analytical laboratory provided chemical analyses. These would typically have included radiometric, fluorometric, and chemical analyses for uranium, plus "impurity analyses."

For the leaching circuit, sulphuric acid was initially made from pyrite[42], an iron sulphide with the chemical formula FeS_2 (iron(II) disulphide). There were two sources for the pyrite: some occurred naturally in the uranium ore and some was mined separately.

For some uranium ores, including the Lorado ore itself, pyrite was contained as an impurity. The Lorado ore, for example, generally contained 4 percent sulphur [145,155]. For these cases, the pyrite was separated from the ore, and blended with pyrite from the mine, in the flotation circuit.

In general, however, the custom uranium ores from other mines did not contain much sulphur. When such custom ores were blended into the mill feed additional pyrite was added, as needed, to the flotation circuit. In order to obtain the pyrite necessary for such additions to the process, it was mined from the separate body of pyrite ore in the Lorado mine, which contained 17 to 20 percent sulphur (and no uranium) [145].

Flotation Circuit. The second-stage filtration cakes from the filtration circuit, described above, were combined with ground pyrite from the mine and processed in a ten-cell, acid-resistant flotation machine (Denver, No. 24) [155]. The flotation was conducted in two stages [15]. In the first stage, graphite was floated and removed while the flotation of pyrite was suppressed (with lime). In the second stage, the slurry was treated with sulphuric acid and copper sulphate and then floated using xanthate as a flotation agent. This provided a pyrite concentrate (30 percent sulphur) that was pumped to the acid plant.

According to Thunaes [153], the free acid level from the flotation circuit tailings was generally not high enough to warrant the return of the solution to the leaching process, so it was simply combined with the other process tailings instead.

---

[42] The mineral pyrite, or iron pyrite, is also known as "fool's gold" due to its gold-like, metallic colour and appearance.

Sulphuric Acid Plant. The sulphur concentrate was roasted in a Dorr Fluo-Solids Roaster [15] - a fluidized-bed reactor in which oxygen was introduced from the bottom to react, at high temperature, with a continuously replenished bed of pyrite particles to produce sulphur dioxide (SO_2) gas. The chemical reaction is:

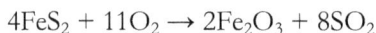

$$4FeS_2 + 11O_2 \rightarrow 2Fe_2O_3 + 8SO_2$$

The SO_2 gas, in turn, was passed through a waste heat boiler and then to the Monsanto 'contact' acid plant proper for conversion to sulphuric acid [15,145,155].

In the Monsanto process, cooled sulphur dioxide gas was passed over vanadium pentoxide (V_2O_5) catalyst in a "converter" to form sulphur trioxide:

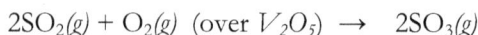

$$2SO_2(g) + O_2(g) \text{ (over } V_2O_5) \rightarrow 2SO_3(g)$$

The sulphur trioxide was then combined with the sulphuric acid[43] in a countercurrent absorption tower to form disulphuric acid (also known as fuming sulphuric acid or oleum).

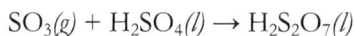

$$SO_3(g) + H_2SO_4(l) \rightarrow H_2S_2O_7(l)$$

The disulphuric acid was then diluted with water to form what was still very concentrated (~93%) sulphuric acid:

$$H_2S_2O_7(l) + H_2O(l) \rightarrow 2\ H_2SO_4(l)$$

The acid plant was designed for the production of up to 45 tonne/day (60 ton/day) of sulphuric acid. The sulfuric acid was stored in steel tanks, from which it was pumped as needed to the leaching circuit.

One major change to the acid plant and process came about as a result of the acid production capacity limitations, noted in Chapter 2, by which it was decided to replace pyrite burning with elemental sulphur-burning. While the existing stockpile of pyrite was being used up, the company initiated the construction of a sulphur melter, sulphur burner, new storage facilities (for the sulphur), a new waste-heat-recovery boiler, and other equipment such as pumps [79,101]. This overhaul was completed in 1959.

---

[43] Sulphur trioxide is not directly combined with water to form sulphuric acid because this is such an exothermic reaction that the product becomes acid vapour and/or mist rather than the liquid.

When the new system was put into operation, solid sulphur was melted in a steam-heated melter, after which the liquid sulphur would have been filtered and pumped to the oil-fired sulphur burner. The liquid sulphur would have spontaneously ignited in response to the high temperature created in the burner and contact with the dried air supplied by an air blower:

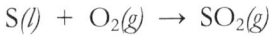

$$S(l) + O_2(g) \rightarrow SO_2(g)$$

The hot gases (at close to 1,000 °C) produced probably contained about ten percent sulphur dioxide. These would have been passed through the waste-heat-recovery boiler, which would have cooled the gas while producing high-pressure steam. The cooled sulphur dioxide-containing gas would then have been passed to the Monsanto acid plant for conversion to sulphuric acid as described above.

Figure 5.10. Sulphur Being Barged to Uranium City, *circa.* **1960. Photographer: Unknown. Courtesy: University of Saskatchewan Archives & Special Collections (F. Walker Collection).**

Other Infrastructure. The principal ore handling and processing circuits were housed in buildings for the scale house, custom ore bins, primary and secondary crushers, and "concentrator." In addition to housing the leaching, filtration, and ion exchange circuits, the concentrator building also contained the mill's laboratory and offices. Other mill-site buildings included the acid plant, powerhouse, auxiliary power plant, booster pump house, acid cooler, chlorate building, chemical storage, warehouse, cement plant, and carpenter's and electrical shops.

The major storage tanks at the mill site included a 145 thousand-litre water tank (32,000 UK gal), two 38 thousand-litre fuel oil storage tanks (10,000 US gal each), an acid storage tank, and a diesel storage tank.

The hot gas from the pyrite roaster in the acid plant was cooled in a waste heat boiler and used to make high-pressure steam. Some of this steam was then reduced, for heating purposes, in a Turbo Generator. This created enough by-product steam to heat all of the buildings on site, except for during the coldest parts of the winter [145].

A diesel-electric power-house provided means to provide power, via two 1200 hp Nordberg Supairthermal diesel generators, steam by way of two 150 hp Dominion Bridge Scotch Dry Back boilers, vacuum via two Canadian Ingersoll-Rand vacuum pumps, and compressed air through a 1,000 cfm (1,700 m^3/h) Canadian Ingersoll-Rand compressor [145,155].

Tailings. The aqueous tailings from the milling processes included stripped effluent from the ion exchange columns, tailings from the flotation cells, calcines from the sulphur-concentrate roaster, and tailings from the precipitation circuit. As such, the tailings contained significant concentrations of uranium, radium, sulphates and chlorides which, of course, were highly acidic (exhibiting pH values in the range of 1.7 to 2.0) [79]. These tailings were simply collected and disposed of, without treatment, by pumping them to a small un-named lake located at the west end of Nero Lake [79].

The volume of tailings produced was so great that within the first year of operation, the tailings overflowed the small lake and flowed into Nero Lake. As noted above, this contaminated the water supply for the milling plant causing numerous operational problems. Rather than change its tailings disposal practices, the company continued to discharge tailings into Nero Lake and built a new water supply line from Beaverlodge Lake, which went into operation in 1958.

Over the mill's operating lifetime an estimated 304 thousand tonnes (335,000 tons) of tailings were released into Nero Lake. At the close of Lorado's operations in 1960, the pH of Nero Lake had dropped to 2.5 near the tailings entry point, and to an average pH of about 3.5 overall [79].

5.4 Mill Evolution.

The original economics for the Lorado mill were calculated based on an average ore grade of 0.28% (as U_3O_8), and an average uranium recovery of 90% (as U_3O_8) [111]. Lorado's Annual Reports show that by the fiscal year ended April 30, 1957, the mill had been in operation for five months, processing about 180 tonnes per day (200 ton/d) of uranium ore and approximately 109 tonnes per day (120 ton/d) of pyrite ore. The processed uranium ore averaged about 0.23% (as U_3O_8), yielding an average uranium recovery of about 93% (as U_3O_8) [111,112,150]. Lorado's pyrite processing in the same period produced an average sulphur recovery of about 92%.

By 1958 Lorado was processing an average of approximately 544 tonnes per day (600 ton/d) and planning for 725 tonnes per day (800 ton/d) [101].

As noted above, the Lorado mill processed ore from its own mine plus ores from other nearby producers, principally:

- Black Bay Uranium Mines Limited,
- Cayzor Athabaska Mines Ltd.,
- Lake Cinch Mines Ltd.,
- National Explorations Limited,
- Rix-Athabasca Uranium Mines Ltd., and
- St. Michael Uranium Mines Limited.

In 1959, even Eldorado Mining and Refining shipped some ore to the Lorado mill. Tremblay [116] records that Consolidated Beta Gamma Mine and Eagle-Ace Mine each shipped in the range of 180 to 270 tonnes (200 to 300 tons) of ore to the Lorado mill in 1959. Table 5.1 (above) shows a broader summary. Beyond these shipments, Lorado also accepted and processed minor amounts of uranium ore from small operations in the region, usually termed "prospector-level" operations, with the intention of "*encouraging exploration and development in the Beaverlodge Area*" [111].

On March 26, 1960, Lorado and Eldorado cancelled the Special Price Contract for the sale of uranium-bearing concentrates, after which all mining and milling operations were halted [114]. The final shipment of uranium (yellowcake) concentrate to Eldorado was made on April 19, 1960.

As shown in Table 5.2, during its operating years the Lorado mill processed approximately 100 thousand tonnes of its own ore plus nearly 400 thousand tonnes of ore from other mines, for a total of over 500 thousand tonnes of ore. From these ores, the mill produced a total of over 1,210 tonnes of uranium (yellowcake) concentrate (as U_3O_8).

Table 5.2. Uranium Ores Processed by the Lorado Mill During its Operating Years [101,111,113,115].

Year Ending	Lorado Ore Milled		Other Mine Ores Milled		Yellowcake Produced (U_3O_8)	
	(tonnes)	(tons)	(tonnes)	(tons)	(tonnes)	(tons)
April 30, 1957	10,900	12,015	31,679	34,920	79	87
April 30, 1958	30,696	33,836	90,969	100,276	440	485
April 30, 1959	29,074	32,049	144,282	159,044	415	457
April 30, 1960	27,242	30,029	103,749	114,364	276	304
Sub-Totals	97,912	107,929	370,679	408,604	1,210	1,334
Total Ore Milled: 468,591 tonnes (516,533 tons)						
Total Yellowcake Produced (as U_3O_8): 1,210 tonnes (1,334 tons)						

6 REFINING AT PORT HOPE

The yellowcake produced from all ores sent to the Lorado mill was refined at Eldorado's refinery in Port Hope, Ontario (Figures 6.1,6.2). Although this refinery had been in operation since 1933, its initial focus was on the production of radium and uranium was simply a by-product. By 1942, the plant's focus had begun to shift to uranium and radium production, and by 1953, the radium production was discontinued, with the plant focusing on just uranium refining [8].

Figure 6.1. The Eldorado refinery, Port Hope, in 1938. Library and Archives Canada (Photo 3375914).

Figure 6.2. The Eldorado refinery, Port Hope, in 1955. Eldorado Mining and Refining Ltd., reference [162].

Although the Port Hope refinery was able to handle the uranium concentrate produced at Eldorado's Port Radium mine, it was realized that additional capacity would be needed in order to handle the additional uranium concentrates produced by the Eldorado Beaverlodge, Gunnar, and Lorado mills, so a new and expandable processing circuit was installed in 1954 and put into operation in 1955 [161,162].

At this point in time, most of Eldorado's purchase contracts for yellowcake were negotiated under a special price formula that specified a specific price per pound of uranium in an acceptable concentrate. The price took into account factors such as estimates of *"daily tonnage and recoverable uranium, and estimated operating, pre-production and capital costs ... predicated on a 5-year period of production during which period the estimated pre-production and capital costs [would] be fully amortized"* [160].

This pricing philosophy was based on the Canadian Government's desire to encourage Canadian companies to find, develop, and operate uranium mines, thus its intent was to have Eldorado mill the ores and refine the concentrates from all non-Eldorado mines and mills on a purely cost-recovery basis [160]. An early problem that arose was that, except for the Eldorado and Gunnar mines, the rest of the Canadian mines (including Lorado) were not likely to be able to remain in production long enough to ensure five years of production before the termination date of March 31, 1962 (beyond which Eldorado did not have contracts to sell its refined product[44]). To accommodate these mines, Eldorado decided to take the risk

86

of increasing the delivery tonnages permitted and to extend the contracts for an additional year [160]. Although the specific contract details seem to have remained classified "Secret" [145], it is likely that Eldorado's contract with Lorado was of this nature.

In Lorado's era (1956 through 1960), the refining process apparently involved dissolving the uranium in nitric acid and then extracting the uranyl nitrate (twice) into ether, then re-extracting it into water (with very large volumes of water), and, finally, heating ("calcining") the purified uranyl nitrate to oxidize the uranium and produce uranium trioxide [67,163]. Accounts vary, with regard to the appearance of the final product, however it seems likely that the original refining process produced "black oxide" (about 95 percent U_3O_8). By the early 1950s, the product was likely "orange powder," a fairly pure form of uranium trioxide (UO_3).

In the early 1950s, the details of the uranium refining processes used in Canada and the U.S. were also held secret [164], although it was known that yellowcake was being processed to yield "orange powder" (uranium trioxide, γ-UO_3). In some refineries, the uranium trioxide was being reduced to "brown powder" (uranium dioxide, UO_2), and then converted to "green salt" (uranium tetrafluoride, UF_4) [164,165] (see Figure 6.3). In the U.S., the uranium tetrafluoride was then converted to uranium metal and/or uranium hexafluoride (UF_6).

Since the late 1950s, however, more information has become publicly available such that several descriptions of the Eldorado refining process are available [1,8,163,164,166]. Only a summary of the process (*circa.* 1955 through 1960) is provided here.

By the time the first yellowcake from the Lorado Mill would have arrived at the Eldorado refinery at Port Hope, in late 1956, the refinery had converted to a simpler solvent extraction process than the one based on ether [164,167]. Research in the U.S. had shown that a better (and safer) approach was to use tributyl phosphate (TBP) in kerosene, and following a commercial demonstration in Ohio in 1954, it was adopted at Eldorado in 1955.

A simplified process diagram for uranium concentrate processing at Eldorado's Port Hope, Ontario refinery is given in Figure 6.4.

---

[44] In 1956, Eldorado's contracts for its refined product were with the U.S. Atomic Energy Commission and the U.K. Atomic Energy Authority. In 1957, the Canadian Government announced that, under certain conditions, it was prepared to sell small quantities of refined uranium to other friendly countries [160].

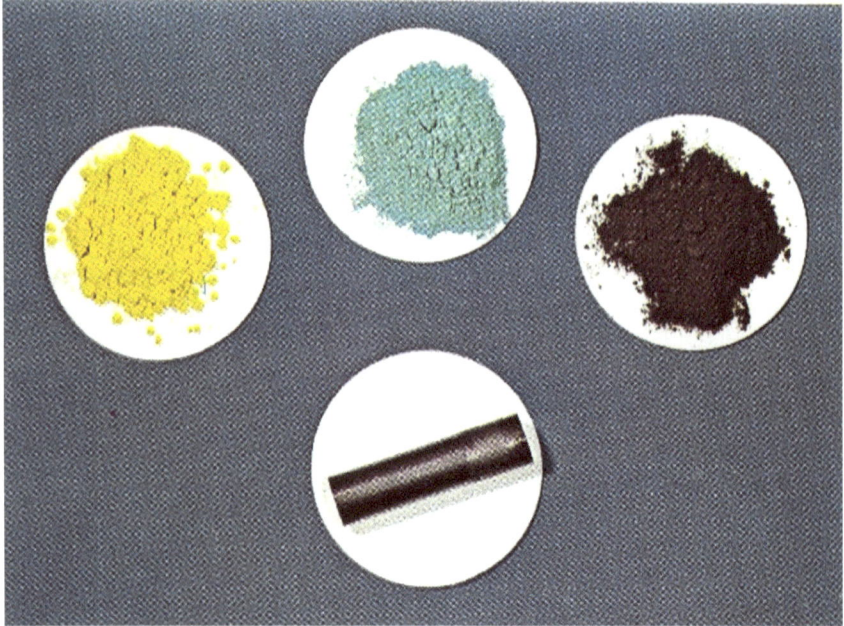

Figure 6.3. Some forms of uranium: yellowcake ($Na_2U_2O_7$, left), green salt (uranium tetrafluoride, UF_4, centre), brown powder (uranium dioxide, UO_2, right), and uranium metal (lower). (Canada Department of Energy, Mines, and Resources [8]).

The following list provides a brief description of each process step:

- *Digestion*. Yellowcake was digested in a series of three digestion tanks. In the first tank, it was dissolved in nitric acid (55 mass% HNO_3) and then cascaded through the next two tanks. Along the way, the process temperature was reduced to 35 °C and the free acid concentration to about 3 molar. A vent system was used to draw off the hot, acidic gases from which nitric acid was recovered and recycled back to the digestion tanks.
- *Solvent extraction*. The aqueous "digest-liquor" was passed to a liquid/liquid extraction column packed with perforated stainless-steel trays (to produce mixing) and pumped in from the top of the column, using a pulse-pump. The organic phase, comprising kerosene containing tributyl phosphate (25 volume% TBP), was introduced into the bottom of the column, also with a pulse-pump, and the uranium was selectively extracted from the aqueous phase into the organic phase as these two phases were mixed, in a counter-current fashion.

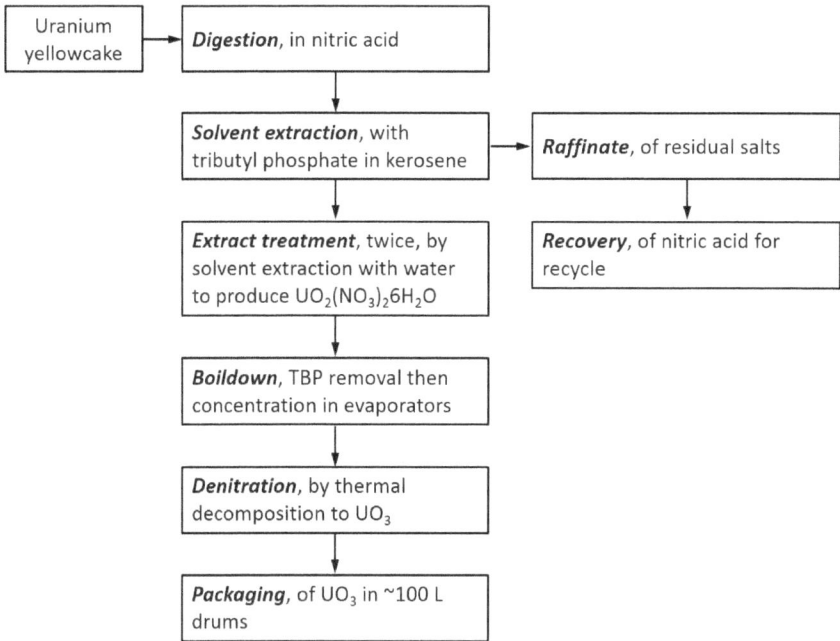

Figure 6.4. Simplified process diagram for uranium yellowcake refining, *circa*. 1958, at Eldorado's Port Hope, Ontario refinery.

Process steps (continued):

- **Raffinate treatment.** The impurity-containing aqueous phase was high in free nitric acid, and a separate plant process was used to recover the nitric acid and recycle it back to the digestion tanks.

- **Extract treatment.** The uranium-containing organic phase was concentrated in a settling tank followed by a centrifuge and then pumped to a "scrub column," which was a second solvent extraction column in which a small amount of water was mixed-in to remove residual impurities and nitric acid. From the scrub column, the organic phased was pumped to the "re-extraction column," which was a third solvent extraction column in which the organic phase was mixed with a large excess of water in order to transfer the uranium back (as uranyl nitrate) to the aqueous phase.

- **Boildown.** The aqueous uranyl nitrate solution was cooled and any entrained TBP was removed by skimming. The concentration of uranyl nitrate was then increased by boiling the solution in large evaporators, yielding a uranyl nitrate hexahydrate product.

- **Denitration**. The uranyl nitrate hexahydrate was then converted to uranium trioxide (UO_3) by thermal decomposition.
- **Packaging**. The uranium trioxide, in granular powder form, was cooled and packed into 100 L (25 gal.) drums for shipping to the U.S. Atomic Energy Commission.

Several other processing options were developed for the Eldorado refinery in the 1950s [8]. By 1950 the refinery was able to produce metal-grade uranium oxide. A "green salt" plant and a metal plant were developed in 1957/58, and Eldorado began producing nuclear-grade uranium metal and metal oxide for the Atomic Energy of Canada Ltd. (AECL) nuclear reactor program. By 1958 the refinery was able to produce nuclear-grade uranium as the metal, as uranium dioxide, (UO_2), or as uranium tetrafluoride (UF_4) [168]. These additional processing and product options are illustrated in Figure 6.5 and briefly described in the list below:

- **Uranium trioxide**. This was the result of the process sequence illustrated above and in Figure 6.4, in which uranyl nitrate hexahydrate was heated to yield uranium trioxide (γ-UO_3, orange oxide) by thermal decomposition.
- **Uranium dioxide**. The uranium trioxide was cast into pellets and, in a moving bed process, heated and contacted with hydrogen gas to reduce the uranium to yield uranium dioxide (UO_2, brown oxide).
- **Uranium tetrafluoride**. The uranium dioxide pellets were placed in a separate vessel in which they were heated and contacted with hydrogen fluoride gas to produce uranium tetrafluoride (UF_4, green salt). Excess acid was removed by scrubbing, and the salt was separated out by filtration.
- **Uranium metal**. The uranium tetrafluoride was mixed with magnesium metal and heated to 1,900 °C, to reduce the uranium to its metal form.

```
┌─────────────────┐
│    Uranium      │
│ Trioxide, UO₃   │
└─────────────────┘
         │
         ▼
┌──────────────────────────────┐
│ Pelletization                │
└──────────────────────────────┘
         │
         ▼
┌──────────────────────────────┐      ┌──────────────────────────────┐
│ Reduction, with hydrogen and │ ───▶ │ Uranium dioxide, UO₂,        │
│ heat in a reaction column    │      │ pellets                      │
└──────────────────────────────┘      └──────────────────────────────┘
         │                                      │
         ▼                                      ▼
┌──────────────────────────────┐      ┌──────────────────────────────┐
│ Fluorination, with hydrogen  │ ───▶ │ Uranium tetrafluoride,       │
│ fluoride and heat in a       │      │ UF₄, powder                  │
│ reaction column              │      │                              │
└──────────────────────────────┘      └──────────────────────────────┘
         │                                      │
         ▼                                      ▼
┌──────────────────────────────┐      ┌──────────────────────────────┐
│ Reduction, with magnesium    │ ───▶ │ Uranium metal, U, ingots     │
│ in a furnace at 1900 °C      │      │                              │
└──────────────────────────────┘      └──────────────────────────────┘
```

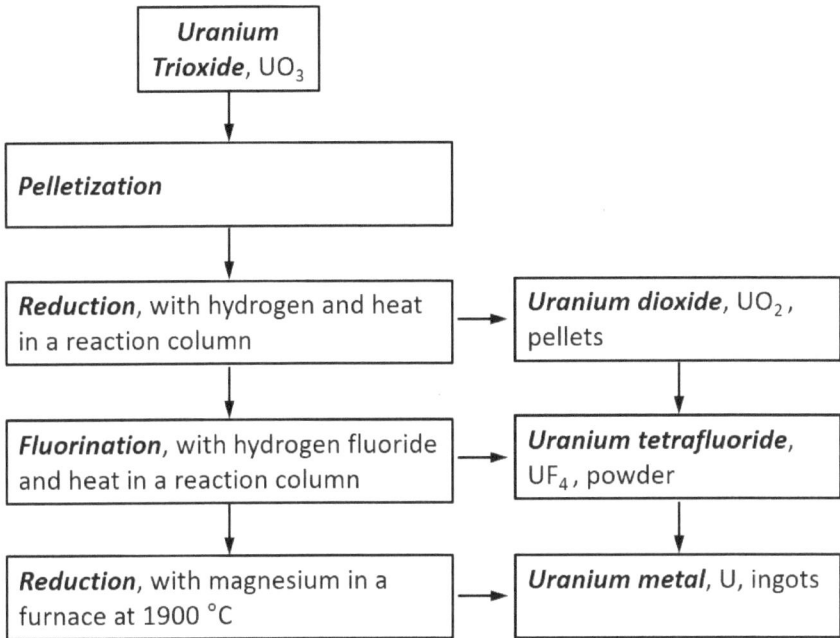

Figure 6.5. Simplified process diagram for uranium products, *circa.* **1958, at Eldorado's Port Hope, Ontario refinery.**

The reason for the diversification in products was the AECL CANDU[45] nuclear reactor development program [168], which, in its early developmental stages, used uranium tetrafluoride as the fuel, and later shifted to using uranium dioxide as the fuel. One of the advantages of the Canadian CANDU design was its ability to use natural uranium as the fuel, without the need for enrichment [169]. Enriched uranium was necessary for export to the United States, however, and for this purpose either the uranium tetrafluoride or uranium dioxide would be re-dissolved, mixed with enriched uranium hexafluoride gas (obtained from the United States), and then reprecipitated.

---

[45] CANada Deuterium natural Uranium reactor.

SCHRAMM, OGILVIE-EVANS, and WILSON

7 ABANDONMENT AND EARLY REMEDIATION

7.1 Closure and Abandonment.

According to Lorado's 1960 Annual Report, all mining and milling operations ended at the end of March 1960, at which time *"[t]he mill crew then proceeded to clean out the circuits"* [114]. The mine property was put on a *"care and maintenance"* basis in April 1961. The company still hoped the uranium market would recover and therefore that the mine and mill shutdown would not have to be permanent. Accordingly [115],

> *"[a]ll surface buildings at the mine and plant site and all uranium treatment equipment of a chemical conversion nature, including the sulphuric acid plant, were 'mothballed' so that they might be rehabilitated, if required."*

The company's hopes of resuming operations at the Lorado site could not have been very high, however, as all equipment that might be used in a metal mining operation elsewhere was dismantled and shipped to Edmonton, Alberta, where it was stored for eventual relocation or sale [115]. Some of this equipment was sold to a British Columbia mining operation in 1961 [115]. The acid plant was sold in 1967 [124].

The company even considered converting its main campsite into a *"high caliber type of fishing resort,"* although it is not clear whether or not this was actually attempted [115].

In 1969, Lorado Uranium Mines Ltd. was acquired by International Mogul Mines Ltd. (see Table 2.2 above).

The mine and mill remained mothballed for the next thirteen years. However, it appears that at some point during this period, the company gave up hope of ever restarting operations at the Lorado site, as the mine shaft, escape raise, and ventilation raise were all cement-sealed, and the head-frame was burned-down.

In 1976, an inspection revealed that two of the mine buildings remained standing, while the rest had burnt and been pushed into a debris pile near the mine shaft [136]. Some of the campsite buildings near the mine were also still standing but in poor condition, including the residences and heating buildings [136]. At this point in time, the mill buildings also remained standing (Figures 7.1 and 7.2).

The mine site was re-staked by a diamond prospector in 1991, but the rock sampling did not yield prospective results, and the claims lapsed in 1997 [136].

Figure 7.1. Aerial view of the Lorado mill site in 1981 (Saskatchewan Energy and Resources).

Figure 7.2. Aerial view of the concentrator building, showing some of the external deterioration (Saskatchewan Energy and Resources, 1982).

Figure 7.3. Aerial view of the Lorado tailings in 1990 (Saskatchewan Energy and Resources).

7.2 Early Remediation.

In early 1982, International Mogul Mines notified the provincial government of its intention to allow their Mineral Surface Leases on the Lorado property to lapse. The company noted, in part, that *"No work has been performed there for at least 25 years. The site was cleared a number of years ago. The shaft was sealed and all but a couple of buildings removed"* [170]. Inspections by the Resource Management Branch identified a number of outstanding issues and cleanup requirements [170], including: the pump-house and bunkhouses to be burned, debris associated with the former shaft-house, machine shop, boiler house, coffee house, cookery, and staff houses to be dozed into depressions and covered with waste rock, and abandoned vehicle bodies and miscellaneous scrap metal debris to be removed from the brush and placed in building foundations or covered over.

The company (by this time, International Mogul had been amalgamated into Conwest Exploration Co.) was allowed to leave the mine-site building foundations in place as long as all protruding metal was sheared off flush with the concrete. Conwest also conducted some re-contouring of the affected landscape. These measures were all completed by the company and the surface leases for the site were cancelled in December 1982 [170].

The mill buildings, having deteriorated over 30 years of abandonment (Figure 7.4), were finally demolished, buried, and the mill site recontoured with clean gravel by Conwest in 1990 [171,172].

Figure 7.4. Remains of the conveyor conduit to the ore storage bins and the main mill building in 1990 (George Bihun, Saskatchewan Environment).

7.3 Status in the Early 2000s.

When KHS Environmental Management Group Ltd. (KHS), working for Saskatchewan Environment, visited the Lorado mine site in 2000 [136] it was found that:

- the mine shaft was well sealed with waste rock, with the winch and headframe foundations still remaining,
- a large quantity of waste rock was evident throughout the mine site, for which gamma radiation levels were reported to range from 0.20 to 1.60 μSv/hr,
- the flow pattern for runoff from the waste rock led into muskeg below the mine site to the north (Figure 7.6), and from there into Beaverlodge Lake. The runoff water was found to have a pH of 2.99,
- there were limited amounts of debris in various locations, including near the remains of the water tower (Figure 7.7),

Figure 7.5. The Lorado mine site in 2000 (Saskatchewan Environment [136]). Note the section of building foundation in the foreground (lower left) and concrete pad in the upper right. The mine shaft was located on a hill overlooking Beaverlodge Lake, which can be seen in the background.

Figure 7.6. Runoff from waste rock at the Lorado mine site in 2000, flowing towards its ultimate destination of Beaverlodge Lake, which can be seen in the background (Saskatchewan Environment [136]).

Figure 7.7. Debris and foundations from the water tower at the Lorado mine site in 2000 (Saskatchewan Environment [136]).

- the fresh water intake weir was still in-place in Beaverlodge Lake, with some debris remaining from the pump house, and
- the campsite near the mine was fairly well cleaned up with no buildings left standing and only a limited amount of remaining debris.

The KHS inspection of 2000 did not include the Lorado mill site, however, a waste disposal site (about 45 m long by 20 m wide) was found in the area [136]. This site was about 1.5 km beyond the Lorado mill site, and just off the road to the Lorado mine. The waste material included ore barrels, piping, and machinery, consistent with what would have been generated during milling operations (see Figure 7.8). Assessment of the waste indicated it was likely created during operations and/or decommissioning of the mill. A number of the barrels were found to contain chemicals and/or fine solids, and gamma radiation levels at this site were reported to range from 0.07 to 5.56 µSv/hr.

Figure 7.8. Waste disposal debris found between the Lorado mine and mill sites in 2000 (Saskatchewan Environment [136]).

The Lorado mill site was assessed by Clifton Associates, working for Saskatchewan Environment, in 2001 [171]. Clifton concluded that, with the shaft and raises having been sealed, the headframe and mine and mill buildings demolished and buried, the remaining concern at the site was with the exposed tailings that could become wind-blown, generate acid, and leach metals into the local ground and surface water. This was no small matter, as it was estimated there were approximately 500,000 m^3 of tailings (58% above ground and 42% below the surface of Nero Lake) covering an area of about 14 ha (35 acres, see Figure 7.9).

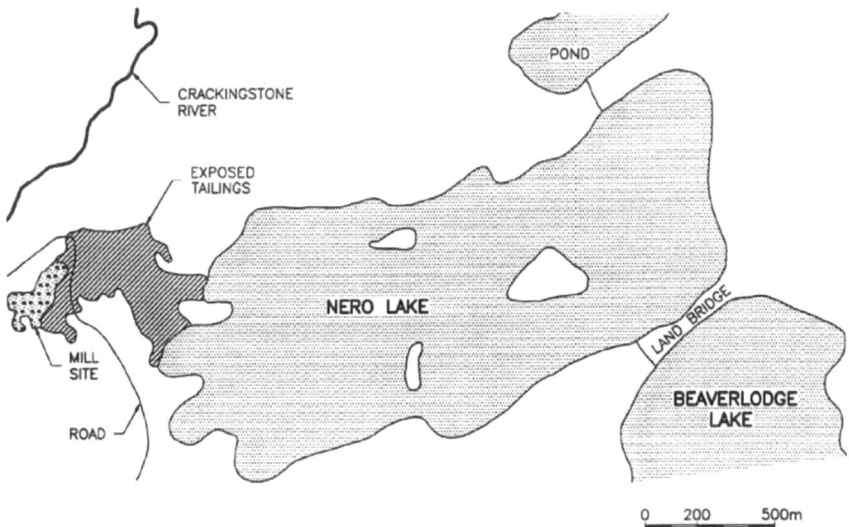

Figure 7.9. Excerpt from a drawing from the Clifton Associates assessment of the Lorado mill site in 2001 (Saskatchewan Environment [171]). The mill site is shown to the left. Note the large expanse of exposed tailings between the mill site and Nero Lake.

The tailings have been characterized as comprising up to 4 m of silt and sand overlying peat, with the larger sand particles occurring at the top of the tailings pile and progressively smaller-sized particles occurring with depth [171]. In addition to the inherently acidic nature of the original tailings, it was discovered that the pyrite in the tailings was continuing to oxidize (generating additional sulphuric acid). For both reasons, there was continual drainage of acidic water into Nero Lake[46].

---

[46] It was also identified that there was some risk of acidic water seeping into the nearby Crackingstone River Valley.

As noted in Section 2.3, Lorado became aware in 1957 that it was seriously damaging the water quality of Nero Lake with continual drainage of acidic water from their tailings into Nero Lake. The company dealt with the impacts on the water supplied to their mill by switching to drawing water from Beaverlodge Lake, but the seepage into Nero Lake continued. It is perhaps not surprising then, that some twenty years later water quality studies of Nero Lake showed that the pH had dropped to about 3.3 (its original value would have been about 7)[47] [171,173]. With no significant fresh-water inlets and only very slow natural buffering mechanisms present [173], there was nothing to counter the continuing seepage of acid into Nero Lake.

By the 1970s it was clear that much of the invertebrate and plankton communities had been destroyed with only some aquatic moss and acid-tolerant invertebrates being able to survive [171,173]. Of these, the aquatic moss was judged to be acting as a sink (i.e., collector) for radionuclides in the lake [173]. Similarly, the exposed tailings were too acidic to support plant life, which increased the degree of wind erosion.

The fish in the lake, of course, had long since been destroyed. Acid mine drainage research of the 1970s clearly demonstrated that while some fish species can survive mildly acidic water, a complete loss of fish occurs by the time the water pH has dropped to 4.5 (see, for example, Cooper and Wagner [174]). At approximately pH 3.3, the Nero Lake water had long been much more acidic than this.

Finally, although Nero Lake and Beaverlodge Lake are separated by a land bridge, it is relatively narrow and porous. As a result, water can flow from Nero Lake into Beaverlodge Lake. Clifton Associates noted in 2002 that this effect could be clearly seen from the air as a plume of water moving from Nero Lake to Beaverlodge Lake [171]. The same plume, at a somewhat later date, is shown in Figures 7.10 and 7.11. It was concluded that the visible plume resulted from the precipitation of dissolved heavy metals, such as aluminum and iron in the Nero Lake water when these metals encountered the slightly alkaline (pH 7.3 to 7.7) water of Beaverlodge Lake. It was found that, in addition to the acid, the Nero Lake water was carrying uranium, radium-236, lead-210, and thorium-230 into Beaverlodge Lake [173]. Ruggles et al. [173] estimated that over the life of the mine and mill some 80 tonnes (8×10^4 kg) of uranium (as U_3O_8) were discharged from the mill into the tailings.

---

[47] For comparison, the pH of pure (neutral) water is 7.0 while apple cider vinegar has an acidic pH of about 3.3.

Figure 7.10. View of Nero Lake (centre), the land bridge, and Hanson Bay of Beaverlodge Lake (lower left). The Lorado tailings area can be seen to the left of centre (Courtesy of Woodland Aerial Photography, *circa*. 2008).

Figure 7.11. Close-up of the land bridge showing a plume extending into Hanson Bay, Beaverlodge Lake (Courtesy of Woodland Aerial Photography, *circa*. 2008).

7.4 Residual Hazards.

In the 1950s it was generally believed that uranium mining and milling posed no particular radiation hazards as long as adequate ventilation was provided, although the need to control dust was recognized [24]. At the time, both uranium mining companies and health authorities were aware that radium posed a health risk. However, information was lacking on specific exposure risks [24]. It wasn't until 1960 that Canadian regulations covering both uranium and radium exposure were brought in by AECB [24].

In the 1950s and 60s, uranium mines in Saskatchewan were not subject to significant pollution control regulations and mine decommissioning was not required [48]. As the 1960s unfolded, public awareness and concerns about radiation and acidity from abandoned and unremediated uranium mines in Canada began to emerge with regard to the Bancroft and Elliot Lake area mine abandonments [24]. The U.S. Government brought in regulations aimed at reducing uranium workers' exposure to radon in 1966 [22].

According to Ruggles *et al.* [173], prior to 1976 little was known about the impacts of uranium mill tailings on natural water bodies. Specific evaluations issued by Environment Canada and Saskatchewan Environment between 1976 and 1978, however, established a clear linkage between tailings contamination and surface water quality degradation.

In subsequent decades the radiation hazards became much better understood and public and government expectations for mine decommissioning, in general, changed substantially. As a result, expectations that abandoned mine buildings be dismantled and removed, and that tailings and waste rock be treated, as necessary, to protect humans and the environment would emerge in later years.

In 2000 the Saskatchewan Government launched an assessment of its northern abandoned mine sites, including uranium mine sites, in order to prioritize them based on public safety and environmental concerns [34]. The assessments related to the Lorado sites were carried out in 2000/01 and 2001/02 [136,171]. Considering both public safety and environmental criteria, each site was evaluated and ranked in terms of risk. A total of 49 sites were evaluated, in two batches. The Lorado mine site, having been mostly remediated, was ranked fairly low for risk at #16 of the 27 sites evaluated in 2000/01 [136]. The Lorado mill site, however, was ranked fairly high for risk at #7, of the 22 sites evaluated in 2001/02 [171].

For the Lorado mill site, and as of 2002 [171], the principal environmental hazard identified was the potential for radionuclides and heavy metals to be transported into other water bodies besides the already severely damaged Nero Lake. The principal public safety hazard identified was the presence of the large area of unconfined tailings that posed a direct gamma-radiation hazard and created the potential for wind-blown tailings to contaminate a larger area.

Uranium mines and mills all have the potential for long-term environmental impacts due to radiological toxicity. The half-lives of the radioactive contaminants of primary concern are:

- uranium-238 at 4.5×10^9 years,
- thorium-234 at 24 days,
- thorium-230 at 7.5×10^4 years,
- radon-222 at 3.8 days, and
- radium-226 at 1.6×10^3 years.

The main radiation hazard to the public was assessed as being direct exposure to gamma radiation, although in general, the Lorado sites exhibited only relatively low gamma radiation levels when surveyed.

In 2004, EnCana installed dust-control fences across the exposed tailings and along the public road, then treated the road itself with calcium chloride for dust control [172]. Also in 2004, EnCana filled a number of sink-holes that had emerged over portions of former mill buildings [172]. In 2005, EnCana rehabilitated the land bridge, which had been found to be in poor condition indicating the potential for failure.

No further cleanup actions were taken over the next decade until SRC began the final remediation work in 2014. Figure 7.12 provides an illustration of the residual pathways for interactions among the human population, the terrestrial environment, and the aquatic environment. These include potential routes for chemicals and radionuclides to leave the source area and to enter the local receiving environment, as well as interactions that may cause physical effects to the environment. The potential pathways included the transport of airborne tailings material and radon, along with the transport of waterborne chemicals and radionuclides through the land bridge [180].

Figure 7.12. Illustration of pathways for interaction among the human population, the terrestrial environment, and the aquatic environment at the Lorado mill site area in 2011. Saskatchewan Research Council [180].

8 FINAL REMEDIATION

8.1 The Modern-Era Need for Additional Remediation.

While the end of the Cold War changed the nature and extent of nuclear developments worldwide, this and the Chernobyl accident of 1986 changed attitudes towards nuclear safety and environmental protection. As a result, many nuclear facilities established since the 1950s became redundant, many others reached the end of their design lives, leaving behind large areas of contaminated facilities and land [175]. According to the International Atomic Energy Agency (IAEA), "Many countries were therefore left with facilities requiring to be decommissioned and/or sites requiring to be remediated" [175][48]. Canada is no exception.

In cases where abandoned mines have left large amounts of tailings and/or waste rock deposited into unsuitable places, the sheer volume and mass of material involved mean that it is usually cost-prohibitive to move them to a more suitable location. This is particularly the case for logistically isolated sites such as Lorado. In such cases, the only practical option may be to allow them to remain in place while conducting remediation measures to minimize ongoing harm to human health and the environment [176].

Canadian federal regulations did not cover uranium mine closures or remediation until May 31, 2000, when the Nuclear Safety and Control Act (NSCA) replaced the Atomic Energy Control Act (AECA). The new legislation was constructed to regulate the complete life-cycle of nuclear activities. As a result, sites like Saskatchewan's Gunnar mine and mill site and the Lorado mill site, that previously existed outside of the jurisdiction of the AECA, did come under the jurisdiction of the NSCA and meant that the Gunnar site had to be licensed by the Canadian Nuclear Safety

---

[48] See the glossary for definitions of the terms "*decommissioning*" and "*environmental remediation*."

Commission (CNSC) [70,177]. CNSC created a Contaminated Lands Evaluation and Assessment Network Program to identify such sites, evaluate them for safety, and make recommendations for the regulatory approach to each site. The Lorado-related area falling under the scope of CNSC licensing was the decommissioned Lorado mill site, the tailings deposit, the area immediately surrounding the tailings deposit, Nero Lake, and Carney Lake (a small lake near the northwest corner of Nero Lake, see Figure 8.1) [180].

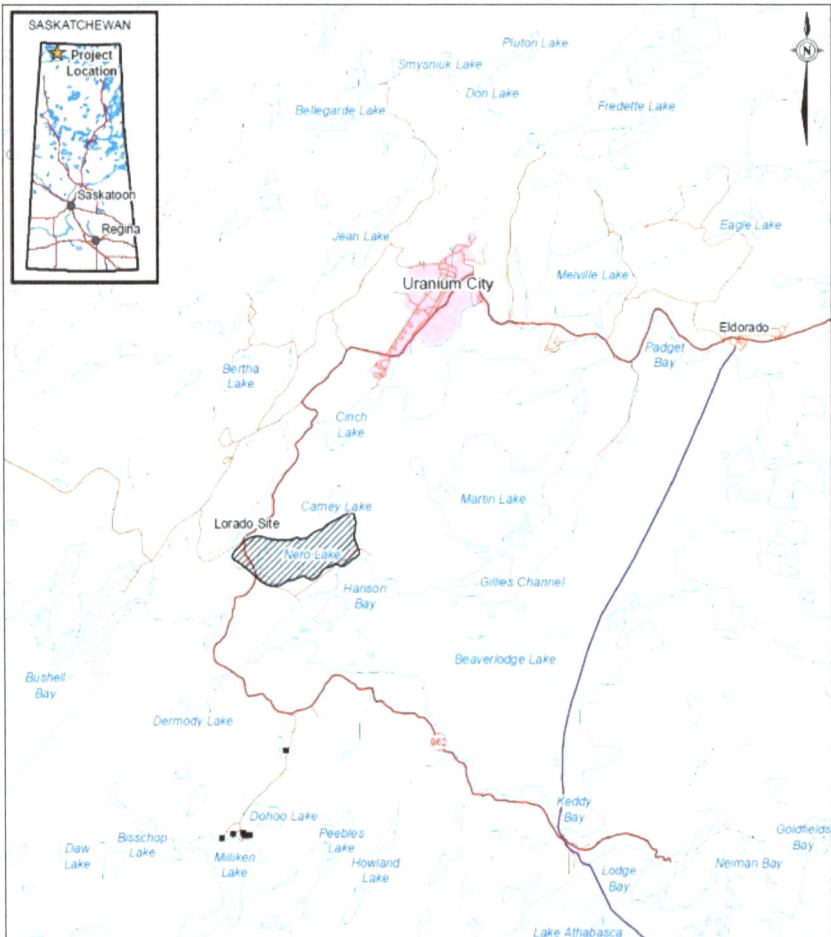

Figure 8.1. Illustration of the Lorado mill site area showing the mill site and tailings area (cross-hatched), Hanson Bay, Beaverlodge Lake (at the southeast corner of the tailings), and Carney Lake (at the northeast corner of the tailings). Saskatchewan Research Council, 2013 [180].

In 2006, the Governments of Saskatchewan and Canada signed a Memorandum of Agreement (MOA)[49] to proceed with the decommissioning and reclamation of 37 Cold War legacy uranium mine and mill sites in Northern Saskatchewan [178,179], which included the remediation of the large Gunnar [70] and Lorado sites, among others. As the property owner, the Government of Saskatchewan had primary operational and legal responsibility for the project. The Saskatchewan Research Council (SRC), a provincial Crown corporation, was contracted as project manager and designated agent to manage and perform the required environmental assessment requirements and rehabilitation activities [179,181]. In addition to the Lorado areas requiring CNSC licensing, SRC's contract with the province also included the Lorado mine site and the waste disposal site that had been found near the mill site (as described in Section 7.3).

The Canadian environmental regulatory regime is complex with both the federal and provincial government legislative frameworks applying. The federal government has authority under federal environmental assessment legislation, fisheries legislation, navigable waters legislation and environmental legislation. The responsible federal departments' oversight is coordinated through the Canadian Environmental Assessment Agency whereas each department has distinct regulatory applications and authorities. In the uranium industry, the federal regulatory framework is made more complex with the CNSC being the ultimate regulatory and licensing authority due to the presence of a "nuclear substance" onsite.

The Saskatchewan provincial government also has approval requirements under provincial environmental assessment (EA) and environmental protection legislation. The provincial government has a joint agreement with the federal government that allows a coordinated provincial/federal approach to environmental assessment.

---

[49] Some of the historical background to the MOA can be found in reference [178].

8.2 Community Engagement.

Numerous local, regional and provincial stakeholders have been interested in the remediation plans and activities, for all of the Uranium City area orphaned and abandoned uranium mines, including the Lorado mine and mill sites. These include the residents of the closest neighbouring communities of Uranium City and Camsell Portage (with a population of about 120), but also a much broader range of stakeholders. As this region is covered by Treaty 8[50], signed in 1899 by the Government of Canada and the First Nations of the Lesser Slave Lake area, the local First Nations have both stakeholder and rights-holder interests.

Given the Lorado sites' location in a remote northern area utilized by First Nations, Métis and Northern residents, communications with these local residents has been of paramount importance and has remained a high priority for SRC throughout its remediation activities. All of the major short-term and long-term site-safety and remedial activities for these sites[51] proposed by SRC include consulting with the local population (see for example [141,182-184]).

SRC also commissioned a traditional knowledge and traditional land-use study so that the environmental assessment and subsequent remediation could be planned and undertaken in the context of traditional uses of the area. This study was conducted by the Prince Albert Grand Council. Similarly, a socio-economic assessment was conducted by SRC on the potential to use biochar as a soil amendment during land reclamation and revegetation, as a possible way to achieve mutually beneficial outcomes for northern communities, as well as the remediation project [185].

A Project Review Committee (PRC) was formed in the early stages to provide a forum that would ensure involvement of each of the impacted communities and enable them to provide direct input on desired remediation endpoints and options, as well as advice on opportunities to maximize the involvement of northern residents in the economic activities emanating from the project. The PRC was established with the assistance of the Prince Albert Grand Council (PAGC) and included elected officials from Prince Albert Grand Council, Fond du Lac First Nation, Black Lake First Nation, Hatchet Lake First Nation, Settlement of Uranium City, Settlement of Camsell Portage, and Hamlet of Stony Rapids. The first

---

[50] Treaty 8 is the largest First Nations treaty by area in Canada and covers a large region of what was formerly the North-West Territories and British Columbia, and includes portions of modern-day northern Alberta, northwest Saskatchewan.

[51] In the specific case of the Lorado sites, EnCana had held four public meetings in Uranium City between 2004 and early 2005. SRC's public meetings were initiated later in 2005, and addressing all 37 Cold War legacy uranium mine and mill sites, including Lorado.

meeting with local Chiefs and Mayors was held in conjunction with a broader town-hall meeting at the Ben McIntyre School in Uranium City in March of 2007, while guidelines for the PRC were finalized and signed by all parties at a meeting in Stoney Rapids in May of 2008[52].

Communications were also established early with the Northern Saskatchewan Environmental Quality Committee (NSEQC). This committee comprises representatives from the northern municipal and First Nation communities that are impacted by northern mining operations in the province, and in particular with the Athabasca Sub-Committee of the NSEQC. The NSEQC monitors uranium mining in Northern Saskatchewan to confirm environmental protection measures and ensure operations are conducted to increase the socio-economic benefits of the surrounding communities. It also serves as a vehicle to enable northerners to learn more relating to uranium mining activities and to see first-hand the environmental protection measures being employed, as well as the socio-economic benefits being gained [186].

Beginning in 2005, SRC held annual public meetings in Uranium City, including discussion of the activities being undertaken at all of the sites. These public consultations included representatives of the NSEQC, the CNSC and Saskatchewan Environment. Other similar consultations have been held periodically in neighbouring Athabasca-basin communities. Tours of the various sites were also provided periodically for members of the PRC and the NSEQC. This range of engagement provided many opportunities for community information exchanges, discussions, and feedback.

Some of the most common questions raised by local community members were:

- What are the impacts of the project?
- What are the remediation options?
- Are there any training opportunities?
- Are there job opportunities?
- How can we actively participate in the remediation?

Other initiatives have included communicating project plans and progress in accessible northern media such as radio, media interviews, flyers, posters, mail-outs, newspapers, and magazines. SRC developed and continues to maintain a *"Project CLEANS"* website (www.saskcleans.ca) [181] for this purpose as well. Information has also been routinely provided

---

[52] The PRC operated for many years and was later wound-up at the request of the committee and replaced with additional, broader, community meetings. The committee had succeeded, credibility, trust, and relationships had been established, and the continuing community meetings have been very effective for all concerned.

to northern media for inclusion in their publications (e.g., *Opportunities North*). A key feature of much of these communications has been translating key project information into the Dene and Cree languages to ensure the information being provided would be broadly accessible [187].

These kinds of engagement activities have been extremely well received by the communities and their leadership:

"... SRC is including communities in the writing of the Environmental Impact Statement ... that type of inclusion has been missing in the past and it's a refreshing change to see ..."

Diane McDonald, Prince Albert Grand Council (2010, [188])

Overall, public support has been very high for the project given that mine sites that had been abandoned for over 40 years are finally being cleaned up [189-191]. The consultation process helped ensure the locally impacted community would be comfortable with the rehabilitation activities and final state of the mine sites. In many cases, advice from the community was adopted into SRC's plans.

An example is the prioritization of the order in which the sites would be remediated. At the first broad-based community meeting in Uranium City in 2007, maps, photographs, and descriptions of each of the 37 sites were presented, posted for viewing, and discussed. Although previous work commissioned by the province had already ranked the sites based on technical public safety and environmental criteria (see Section 7.4 and references [34,136,171]), SRC asked residents of the Athabasca Basin communities to provide their own input as well, based on their perceptions of the accessibility and risks to their families and other potential visitors to the sites. With this input SRC modified some of its priorities, such as by moving quickly to address concerns relating to wind-blown tailings at the Lorado site. This involved upgrading the wind-barrier fencing on the tailings, and applying dust suppressant, to the tailings and road. The upgraded fencing and dust suppression treatments were maintained/repeated in subsequent years until the tailings were finally properly remediated [180]. Overall, the consultation process contributed to a sense by the communities that their views and concerns were being heard and were having an impact.

In another example, based on advice from the advisory committee and community meetings, SRC revised its procurement process for contract work on the project and also arranged training programs, all with the aim of promoting bidding from and hiring of northern contractors and northern employees [37,70,189-191].

One of the most significant challenges has been meeting local expectations for economic benefits given the limited project funds available. Efforts have been dedicated to training local communities and Aboriginal entities in such aspects as the tendering process, safety practices, equipment operation, and so on. The project work was compartmentalized to allow local participation in a variety of tasks including light equipment operation, as well as tenders developed that encourage the use of local workforces. Efforts were also undertaken to allocate project funds to hands-on training opportunities for work occurring outside the tendering process.

When the Gunnar mine and mill site remediation project was approaching the 2010/2011 demolition phase, described elsewhere [70], both the PRC and the communities were engaged in discussions as to how to maximize local employment during this phase. Community liaison positions were created to coordinate employment and training opportunities for individuals at the community level. Programs were developed and put in place to train local communities and Aboriginal entities in the tendering process, safety practices, etc., to maximize the ability of local companies to bid on work and to maximize the number of northern residents qualified to work on the demolition. Such programs have made a significant difference, not only for the Gunnar and Lorado remediation projects but for all the rest that have and will follow.

Key goals for all 37 sites have been to ensure that at the end of these remediation initiatives all stakeholders are satisfied that the sites pose no significant dangers to public health and safety, are not a source of ongoing pollution or instability, and allow for productive use of the land similar to its original (pre-mining era) uses, or at least for acceptable alternative uses.

Figure 8.2. Aerial photo of the Lorado tailings area. The mill was originally to the lower-right of the tailings. Nero Lake is shown in the centre, with Hanson Bay and Beaverlodge Lake to the upper-right. (Courtesy of Woodland Aerial Photography, *circa*. 2008).

8.3 Final Remediation.

Under Project CLEANS, the final Lorado sites remediation project consisted of:

- Final remediation of the Lorado mine site,
- Clean-up of the waste disposal site that was found near the mill site,
- Improving the water quality of Nero Lake,
- Covering and revegetating the surface tailings in-place using a capillary break type soil cover, and
- Appropriate monitoring during and after rehabilitation.

The principal remaining issue at the Lorado mine site concerned improving the coverings of the entrances to the underground workings, namely the main shaft, vent raise, and escape raise. Each of these was cleared of adjacent bedrock at the mine site, surveyed, and recovered with temporary steel plate covers (see Figure 8.3). As of the time of writing, design work was underway for permanent, stainless-steel caps which will be installed during future remediation activities.

Figure 8.3. Lorado mine shaft covered with a temporary steel plate in 2017 (Saskatchewan Research Council).

Beyond covering the entrances to the underground workings, some mine site debris and hazardous waste materials remain stockpiled in several centralized locations and (as of the time of writing) are awaiting proper management during future remediation activities. The Lorado mine is one of the few sites in the Uranium City area that has acid rock drainage issues associated with highly acidic, and possibly high metals-content, solution emanating from flowing boreholes.

In addition to the wooden stave water pipeline (running from Beaverlodge Lake to the mill) and the waste disposal site that had been found in the vicinity of the mill, several other nearby debris dumps, waste rock piles, concrete foundations, and sulphur deposits were discovered (Figure 8.4). Some of these contained leftover mill chemicals and fuels. All of these were included in the mill site remediation work and were covered in a similar fashion to the tailings (see Section 8.5).

Figure 8.4. A debris dump near the Lorado mill and tailings area. (Author photo, 2015).

An early result of SRC's 2005 community consultations related to local concerns regarding wind-blown tailings. As noted above this resulted in early actions to upgrade and maintain wind-barrier fencing on the tailings, and intervals of applying dust suppressant, to the tailings and road. These were temporary measures aimed at reducing the potential for wind-blown tailings to contaminate a larger area. Figure 8.5 shows the upgraded fencing.

Figure 8.5. Aerial view of the Lorado tailings showing the upgraded wind-fencing mine (appearing here as red diagonal lines) *circa.* **2008 (Courtesy of Woodland Aerial Photography).**

The dust suppressant chosen was a formulated commercial product having the tradename EK35® (Midwest Industrial Supply Inc.). EK35 is a proprietary formulation of iso-alkanes, paraffinic alkanes, and long-chain carboxylic acid surfactants [192-194]. This formulation is designed to coat and bind the smaller particles together, preventing their erosion and transport by wind. Figure 8.6 shows the dust suppressant being applied to the tailings area and adjacent public roadway in 2011.

The remediation of Nero Lake and the surface tailings themselves were much more involved, and of course, they were not independent of each other. The water quality of Nero Lake was still being affected by the tailings. Runoff from the tailings area was dissolving the salt precipitate on the tailings surface and transporting it to Nero Lake, thus contributing both acidity and dissolved metals. The Nero Lake restoration and covering of the exposed tailings are described in Sections 8.4 and 8.5.

Figure 8.6. Dust suppressant being applied to the public roadway (upper) and to the Lorado tailings (lower) in 2011 (Saskatchewan Research Council).

8.4 Restoration of Nero Lake.

Nero Lake is approximately 2.2 km long by about 1 km wide. It has an estimated volume of 11.0 x 10^6 m^3 (11.0 x 10^9 l) and an estimated area of 1.72 km^2 (172 ha, 425 acres). The mean depth of the lake is 5.8 m and the maximum depth is close to 13 m [172]. Golder Associates have estimated that about 70% of the lake volume is contained in the upper 5 m [172].

Nero Lake had been so drastically altered by the tailings from Lorado's milling operations over the years that by the 1970s, if not sooner, it could no longer be considered to be a natural waterbody.

In the 2000s, the lake was still quite acidic (pH 4.0) and with elevated concentrations of aluminum, copper, lead, manganese, nickel, titanium, and zinc relative to nearby reference areas (Milliken Lake and Keddy Bay) [180]. The lake-bed tailings were found to cover approximately 40% of the bottom of Nero Lake and there was enough acid inventory remaining in the surface tailings deposit to keep on supplying acidity and dissolved metals to the lake for the long term, meaning hundreds of years [180].

The goals of the remediation of Nero Lake were not to completely re-establish it as a natural waterbody, but rather to improve its ecosystem health and to stop the flow of acidic water into Hanson Bay, Beaverlodge Lake.

Having evaluated various options, it was ultimately decided that the best approach was to cover the surface tailings in order to minimize further contamination and to directly treat the lake water to restore[53] (approximately) its original pH [180]. Water quality modelling showed that the lakebed tailings would only have a minor effect on predicted long-term lake-water quality. It was therefore judged that covering of the lakebed tailings was not necessary [180].

Given the remote location of the lake, the base selected for the addition was CaO (lime), having considered purchase cost, mass and transportation costs, and hazard assessment.

A plan was then developed for the batch treatment of Nero Lake with lime (calcium oxide, CaO) to increase its pH to approximately 7, which would in turn cause *in situ* precipitation and settling of aluminum hydroxide, thus containing the aluminum and preventing its further transport into Hanson Bay of Beaverlodge Lake [180]. Lime reacts readily with water to produce calcium hydroxide (slaked lime), which in turn reacts quickly to neutralize excess acid.

---

[53] Even with the surface tailings dealt-with, it was estimated that recovery of Nero Lake's previous pH balance would take 50 to 100 years, so it was decided to directly treat the lake water as well.

$$CaO\ (s) + H_2O \quad \rightarrow \quad Ca(OH)_2$$

Once the solution pH reaches 5 to 8, dissolved aluminum precipitates,

$$Al^{+3} + 3OH^- \quad \rightarrow \quad Al(OH)_3 \downarrow$$

At higher pH, and depending on the oxidizing nature of the solution, the aluminum may form complex precipitates with magnesium, and eventually dissolved manganese also precipitates as manganese oxide and/or hydroxides.

A treatment process for the lake was developed and pilot tested in 2013. This involved evaluating the lime addition requirement to neutralize the excess acid, precipitate most of the dissolved aluminum and manganese, and provide some residual alkalinity to provide chemical stability [195].

Although it would have been ideal to treat all of the water in the lake directly, this was judged to be too expensive given the very large process and pumping equipment that would have been required. The pilot testing work, however, had demonstrated that it was feasible to treat only a fraction of the water with a higher dosage of lime. Based on the pilot testing results, scale-up, and cost-benefit analyses it was decided to treat 20% of the total volume (i.e., 2.2 million m^3) of the lake with 400 tonnes of lime (Figure 8.7), and then allow natural mixing and equilibration to do the rest *in situ* [195].

The process for the lake treatment involved withdrawing water from Nero Lake, mixing it with the lime to form a slurry (slaked lime), and transferring it to agitated storage tanks where it was further diluted to about 10 mass %. The diluted slurry was pumped from the tanks to the middle of the lake, where it was discharged (see Figures 8.8 and 8.9).

It was decided to proceed with a 400-tonne treatment in 2014, allow time for natural mixing and equilibration, and then evaluate the results at the beginning of the next field season and then make a further addition of lime, if necessary. This was done, with 12% of the lake being directly treated, and the rest indirectly treated. The treatment was successful as by October 2014, the pH of the lake had been brought into the range of 8 to 9.

When the lake condition was checked at multiple stations in June 2015, all samples showed a pH of 7.3. When re-checked in September of the same year, the water pH was 7.1. The alkalinity of the water throughout this period was approximately 28 mg/l, greater than the target value of 5 mg/l. With the lake water in essentially neutral condition, with some buffer capacity restored, and given that geochemical modelling predicted an equilibrium pH of about 6.7, it was decided that there was no need for further lime treatment (although monitoring of the lake water quality has continued).

Figure 8.7. Unloading of 400 Super Sack® bags of dry lime. Each sack held about 1 m^3 and weighed about 1 tonne. (Author photo, 2014).

Figure 8.8. Making the initial lime slurry (left) and pumping it from agitated storage tanks to the lake (right). (Author photos, 2014).

Figure 8.9. Aerial of the water treatment infrastructure in 2014. Note the two sets of pipelines leading out into the lake (lower right). (Saskatchewan Research Council).

As of the time of writing (2018), all monitoring results show that the one-time treatment of Nero Lake was successful. In addition to meeting the threshold levels, many of the parameters of potential concern (i.e., aluminum, manganese, uranium, and zinc concentrations) decreased to below CCME[54] freshwater quality guidelines in Nero Lake.

---

[54] Canadian Council of Ministers of the Environment (CCME) Canadian Water Quality Guidelines.

8.5 Remediation of the Lorado Tailings.

The volume of tailings on land was estimated at 177,000 m³, with oxidized tailings representing an estimated 52,000 m³. The tailings are composed of an oxidized material in the upper regions of the tailings and unoxidized material in the deeper regions of tailings. The surface tailings deposit was up to 4 m thick, consisting of lower silt tailings overlain by upper sand tailings. The upper portion had been oxidized to depths ranging from 0.1 to 2 m across the tailings area.

Having evaluated various options [180], it was ultimately decided that the best approach was to excavate all debris, impacted soil on the road adjacent to the tailings deposit, and on the mill ridge, and other impacted areas, place the excavated material on the surface tailings, fill and recontour the excavated areas, and then cover the surface tailings in-place using an engineered, capillary-break type, soil cover. The soil cover would serve to reduce gamma radiation and radon emissions, eliminate tailings dust, reduce tailings acid generation and prevent the formation of efflorescent salts[55] on the surface (Figure 8.11), and block uptake of tailings pore fluid into vegetation. The soil cover also had to cover any impacted soils and vegetation adjacent to the tailings that may have been contaminated.

Figure 8.10. Aerial view of the Lorado tailings before covering (Saskatchewan Research Council, 2011).

---

[55] See the glossary for a definition of the term "*efflorescent salts.*"

Figure 8.11. Efflorescent salt residues on the Lorado tailings before covering [180] (Saskatchewan Research Council, *circa*. 2009).

Figure 8.12. One of many radiation warning signs, in English and Dene, at the Lorado tailings area before covering the tailings (Author photo, 2014).

Furthermore, it was also decided to cover the submerged tailings, from those at the predicted low-water level in Nero Lake to a depth of 3 m, since such tailings had the potential either to be exposed at the low water mark or to have insufficient water cover to mitigate the identified risks. On the submerged tailings, only a sand cover was required. Tailings with more than a 3 m water cover were left in place.

The capillary-break cover system involved 0.25 m of till material over 1 m of specifically sized, unsaturated sand material (Figure 8.13). In this system, the sand layer acts as a capillary break between the tailings and the till material. This system limits water transfer across the tailings-cover interface while maintaining the active zone of infiltration and evaporation within the upper till layer. At the same time, the till layer limits the infiltration of meteoric water[56] into the tailings material and supports vegetation growth for long-term cover stability.

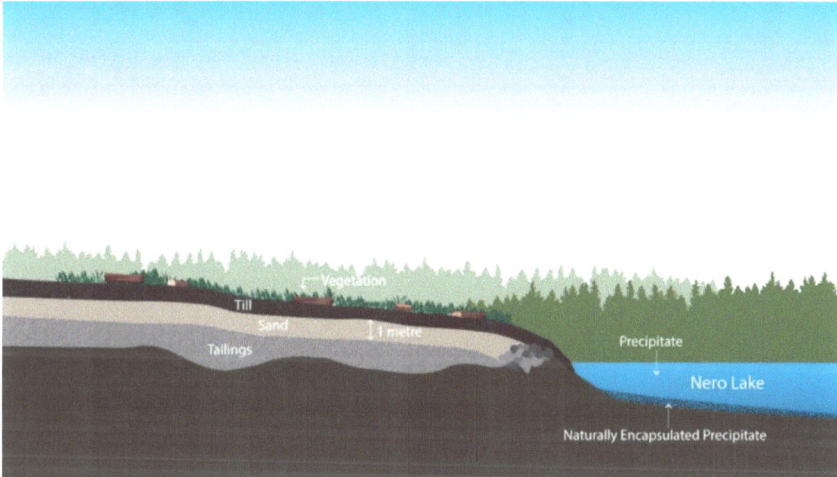

Figure 8.13. Illustration of the engineered capillary-break cover system for the Lorado tailings (Saskatchewan Research Council, 2016 [191]).

Figures 8.14 and 8.15 show stages in the covering of the surface tailings.

Overall, the covers used about 343,900 m^3 of borrow material to cover the tailings, of which an estimated 93,400 m^3 was till and 250,500 m^3 was sand [191]. The cover system was also designed to ensure that it is free draining and contoured to shed water towards one of two engineered drainage ditches, or directly to Nero Lake.

---

[56] Water from precipitation, such as rain and snow. Meteoric water also includes surface water from lakes and rivers, which indirectly come from precipitation.

Figure 8.14. Aerial view of the Lorado tailings covering operation underway in 2014 (Saskatchewan Research Council).

Figure 8.15. Aerial view of the Lorado tailings covering operation in 2015, showing the two engineered drainage ditches under construction (Saskatchewan Research Council).

Subsequent to the completion of the tailings cover 2016, regular monitoring has confirmed that the remediation objectives for gamma radiation were achieved. For example, the average dose rate from gamma exposure is less than 1.2 μSv/h and average radon concentrations vary from below the detection limit (< 20 Bq/m^3) to 40 Bq/m^3.

Revegetation. In the decades since Lorado's closure, several studies related to the potential for revegetation in the region have been conducted [196]. SRC carried out additional revegetation research, consulted with local elders, and set-up three trial plots on a completed portion of the tailings cover to test seeding and fertilizer rates that had been developed for this area. The results were used to adjust the final seeding and fertilizing rates on the cover. Ultimately a native plant seed mixture was designed, consisting of six kinds of grass and three forbs (herbaceous flowering plants).

Terrestrial ecosystem monitoring results were used to identify areas where natural recovery would likely be successful, as well as sensitive areas requiring seeding to enhance erosion control and promote vegetation establishment. In total, approximately 16 ha of the site were left for natural recovery, and approximately 40 ha were seeded with fertilizer application between 2015 and 2017. Figures 8.16 and 8.17 show stages in the emergence of the new vegetation.

Figure 8.16. New vegetation emerging from the Lorado tailings cover in August 2017 (Saskatchewan Research Council).

Figure 8.17. New vegetation emerging from the Lorado tailings cover in June 2018 (Author photo).

8.6 Transfer to "Institutional Control."

At the time of writing, the Lorado remediation is substantially complete. Other than a few pending actions, mentioned in earlier sections of this chapter, the Lorado remediation is expected to meet the design requirements without active maintenance. It is SRC's intention to ultimately transfer the remediated Lorado mine and mill sites into the Saskatchewan Government's Institutional Control Program (ICP), which falls under The Reclaimed Industrial Sites Act (2007), and for which the province has Reclaimed Industrial Sites Regulations [178]. The endpoint criteria for the Lorado remediation project were developed with this process in mind, with a view to requiring minimal maintenance over the very long term. The endpoints are [180]:

- Block all pathways for human exposure to tailings (gamma radiation and radon exposure), including exposures by inhalation of tailings dust, ingestion of vegetation, and ingestion of meat via food chain uptake into animals,
- Block all pathways for wildlife exposure to tailings, including exposures by inhalation of tailings dust, ingestion of efflorescent salts, ingestion of vegetation, and ingestion of soil invertebrates, and
- Prevent additional formation of precipitate in Hanson Bay by blocking the Nero Lake exposure pathway (by improving the Nero Lake water quality).

To this end, the specific remediation activities have been [180]:

- Covering the surface tailings using an engineered cover system that incorporates a capillary break type soil cover, including tailings within the immediate beach area and the adjacent affected soil and vegetation (including that on the mill ridge and road),
- Excavation of any debris or impacted soil on the mill ridge, the road, or any other area associated with the Lorado site, placing the material with the tailings mass, and grading the excavated areas to a safe and stable condition prior to placing the soil cover,
- Treatment of Nero Lake water using an *in situ* batch treatment process to increase the lake pH, and to prevent the formation of additional aluminum precipitate within Hanson Bay, and
- Implementing a performance monitoring program to confirm the physical and environmental stability of the soil cover and the Nero Lake water quality during and after the decommissioning period.

The endpoints allow for traditional and non-traditional land use including, for example, plant collecting, travel, short-term occupancy (such as camping), hunting and fishing at the Lorado site.

9 RETROSPECTIVE

Like other uranium mines of its time, Lorado's mine and mill boomed briefly, when the industry was "hot," and collapsed when the bulk of the ore ran out. Such Cold-War-era boom and bust cycles were repeated in Canada, near Uranium City, Saskatchewan [39,45,46,48,70] and Elliot Lake, Ontario [24,197,198], and also in the United States, in such locations as Uravan, Colorado; Moab, Utah; Jeffrey City, Wyoming; and Grants, New Mexico [5].

In the case of Lorado, like so many of the others, a natural resource was developed and exploited benefitting a local community, a region, and even a country, providing economic and national security benefits. On the other hand, these sites left behind a legacy of environmental disruption and damage, human and animal health risks, along with fearsome clean-up costs.

The Lorado mill's total uranium production, including that from the processing of ores from nearby mines, came to an estimated 1,210 tonnes of U_3O_8. This would have been worth approximately $22 million Canadian dollars in 1960, which would be worth close to $185 million in 2018 Canadian dollars[57]. A portion of Lorado's uranium sales would have been for its own mined ore, a portion to the other local mining operations that fed Lorado's mill, and Eldorado may also have made money on the (nominally cost-recovery basis) refining of the yellow cake at its Port Hope, Ontario refinery.

Based on the experiences of the nearby Gunnar Mines operation [70], it seems likely the total uranium royalties paid to the Saskatchewan and Canadian governments would have been in the order of $1 million dollars

---

[57] $1 Canadian in 1960 would be worth $8.49 Canadian in 2018 according to "Inflation Calculator," http://inflationcalculator.ca/.

each (in 1960 dollars).

In addition, the principal driving purpose of the uranium exploration wave that led to the development and operations of Lorado's mine and mill was to find and develop uranium fuel for strategic military purposes during the cold-war era (see Section 1.4). This was surely a success as the uranium produced at Lorado would have contributed significantly to the cold-war effort. The construction of the Lorado custom mill ensured the success of other, smaller mines in the area that may otherwise have failed. Taken together, the 16 Beaverlodge area uranium mines contributed substantially to this effort in the Atomic Age and Cold War Eras (Section 1.4). How much uranium from the Lorado operations ever found its way into nuclear weapons is either unknown or classified, however, most of it would likely be contributed to such weapons, or to the reserves for such weapons, or both. Lorado's contributions to atomic bomb research and development were later applied to peaceful uses, such as nuclear medicine and nuclear power. Whether any, or all, of these uses of uranium amount to a net positive or negative benefit for society overall, continues to be publicly debated.

The other Lorado legacy, of course, is the substantial clean-up efforts that were left to its successor companies, including International Mogul, Conwest, and EnCana, and to the Province of Saskatchewan and SRC. Although the cost of SRC's work is known, expenditures of the earlier companies are not, so the total remediation cost for Lorado is not completely known.

With the remediation of the Lorado mine and mill sites now substantially completed, Lorado joins a short world-wide list of uranium mines have been completely or substantially remediated. This includes the original Shinkolobwe mine, the Democratic Republic of the Congo, in the early 2000s [199] and only a handful of others. According to the International Atomic Energy Agency, many parts of the world have experienced large delays in advancing the decommissioning and remediation of nuclear sites, and for a variety of reasons[58] [200]. Only a few of Canada's uranium mines have been substantially, let alone completely remediated:

- The Cluff Lake mine in Saskatchewan was remediated as of 2013 by AREVA Resources Canada Inc. [201],
- The Agnew Lake mine in Ontario was remediated as of the early 1990s by Kerr Addison Mines [201],
- The mines in the Bancroft area of Ontario (Dyno, Bicroft and

---

[58] Including issues related to national policies and frameworks (or the absence thereof), financing, availability of technology and/or infrastructure, stakeholders, and/or politics – see reference [195].

Madawaska) were remediated in the 1980s and 1990s [201],

- Of the 12 mines in the Elliot Lake - Blind River area of Ontario, five of the sites had been decommissioned by about 2002, and all of the rest have been decommissioned since that time. At the present time, all of these mine sites have been substantially remediated, with their mine features capped or blocked, facility structures demolished, and the sites landscaped and revegetated (although still requiring some ongoing water treatment) [197,201],

- The Rayrock mine in the Northwest Territories was remediated by the Government of Canada in 1996 [201],

- The Port Radium mine, also in the Northwest Territories, was partially decommissioned in 1984 and fully remediated by the Government of Canada by 2009 [201,202],

- The Gunnar mine, plus some 35 smaller "satellite" mines in Northern Saskatchewan are currently being remediated by the Saskatchewan Research Council (SRC) [181], and

- The Beaverlodge mine in Saskatchewan comprises 62 licensed properties of which some have been fully remediated while others are still being remediated by Cameco Inc. on behalf of Canada Eldor.

Although it is inappropriate to compare the cleanup of the Lorado sites to those of much larger mines, the latter do provide illustrations of how expensive the remediation of such legacy hazards from the past can be. For example:

- In Canada, the nearby Gunnar uranium mine produced over 5 million tonnes of uranium ore, from which the Gunnar mill produced over 8 million kilograms of yellowcake (U_3O_8) between 1955 and 1964. The value of the uranium produced by Gunnar is about $1,100 million (in 2016 Canadian dollars), and the latest public estimate of the total remediation cost is $250 million (also in 2016 Canadian dollars) [203,204]. The remediation itself is still underway [70].

- In the United States, the Mi Vida uranium mine in Colorado produced 12 million pounds of uranium ore during its operating life, "*enough to make at least eighteen atomic bombs*" [22]. The associated Utex mill remediation project has been estimated at U.S.$400 million [22].

- In Australia, the Rum Jungle was a uranium deposit in the Northern Territory, Australia. Discovered in 1949, a mine and mill were constructed in 1952 that operated from 1953 to 1971. The initial 10-year project produced about 3.2 million pounds of uranium oxide [205]. Upon closure, the Australian government decided not to rehabilitate the mine site, as a result of which acid and metals leached into the nearby East Finniss River for many years. In addition, the abandoned open-pit mine was converted to a lake, which also became contaminated. After mining, the area suffered elevated gamma-ray radiation, alpha-ray emitting radioactive dust, and significant radon concentrations in air. Successive attempts to clean up the Rum Jungle site were made in 1977, 1983, 1990, and again in 2009, spending over A$25.7 million. In 2003, a government survey of the tailings piles at Rum Jungle found that capping which was supposed to help contain this radioactive waste for at least 100 years, had failed in less than 20 years. It has been estimated the final remediation could cost an additional A$100-200 million. Reference [206].

These stories illustrate that, when not properly planned-for from the beginning, the remediation phase of such industrial development can end-up costing as much or more than the value of the original extracted resource. A key lesson is that mine and mill remediation and reclamation are best considered, planned-for, and budgeted-for at the beginning (before mining ever begins), as part of a comprehensive, full-cycle (sometimes referred to as "cradle-to-grave") approach to uranium development.

10 GLOSSARY

25   A World War II-era code word for uranium-235. *See* Atomic Code Words.

42-17 grade Z A World War II-era code word for uranium oxide. *See* Atomic Code Words.

Adit   A horizontal passage, driven from the surface, for the purpose of mining or dewatering in a mine.

AECB   *See* Atomic Energy Control Board.

AECL   *See* Atomic Energy of Canada Ltd.

Atomic Code Words During the World War II-era atomic power research and uranium production were conducted in secrecy. In communications among partners Canada, the U.S., and U.K. code words were used to refer to materials such as uranium oxide (42-17 grade Z), uranium-235 (25), and heavy water (polymer) [24].

Atomic Energy
Control Board (AECB) An entity created in August 1946, under the Atomic Energy Control Act, to control and supervise "*the development, application and use of atomic energy*" [24]. The Board had a wide regulatory authority that spanned research, mining, production, transportation, and use of prescribed "atomic substances" [24]. AECB was superseded by the Canadian Nuclear Safety Commission in 2000.

Atomic Energy of
Canada Ltd. (AECL) A Crown Corporation created in 1952 to assume responsibility for the nuclear research program formerly conducted by the National Research Council of Canada.

Atomic Energy
Worker The 1960 AECB regulations defined for the first time the concept of workers in jobs that could cause them to be exposed to nuclear radiation. AECB also regulated the maximum amounts of radiation to which such a worker could be allowed to become exposed. The AECB regulations also defined the maximum amounts of ionizing radiation to which a member of the general public could be allowed to become exposed, at $1/10^{th}$ of the amount for an atomic energy worker. In modern practice, the term has become "*Nuclear Energy Worker (NEW)*" and is defined by the Canadian Nuclear Safety Commission.

Atomic Pile The first nuclear reactor cores contained a "pile" of layers of uranium pellets alternating with graphite bricks.

Black Oxide An impure form of uranium trioxide (which was about 95 percent U_3O_8). *See* Refined Uranium.

Bright Orange
Powder A fairly pure form of uranium trioxide, UO_3. *See* Refined Uranium.

Brown Oxide Uranium dioxide, UO_2. *See* Refined Uranium.

Cage   A cage-like elevator car, suspended from a hoist on steel wire rope and used to transport miners and equipment up and down an underground mine shaft. Also called a Mine Cage. At the Nicholson Mine, the cage compartments were also used to hoist ore and waste rock. For this, tram cars were simply rolled into and out of, one or both of the cages. *See also* Skip.

Canadian Nuclear Safety
Commission   (CNSC) Canada's modern-day regulator, which regulates the use of nuclear energy and materials to protect health, safety, security and the environment. CNSC was established in 2000 to replace the former Atomic Energy Control Board.

CANDU   CANada Deuterium natural Uranium reactor. Canada's main commercial nuclear power reactor design, which uses pressurized heavy water as the moderator. The first commercial CANDU reactors were developed in the 1950s and 1960s. The acronym CANDU-PHW is sometimes used to distinguish this design from other experimental CANDU designs.

CANDU-PHW
     See CANDU.

CCME Guidelines
     Canadian Council of Ministers of the Environment (CCME) Canadian Water Quality Guidelines.

CNSC   *See* Canadian Nuclear Safety Commission.

Cilgel   A "semi-gelatin dynamite," typically comprising 50% nitroglycerin mixed with ammonium nitrate and sodium nitrate, and a hydrocarbon like tar (to make it waterproof). *See also* Driftite, Forcite.

Cobbing   The process of breaking up (usually blasted) ore in order to separate ore-grade material from waste rock. Hand-cobbing refers to doing this by hand, usually using a hammer.

Core	Cylindrical pieces of rock of various lengths that are cut and brought to surface through diamond drilling.

**Dark Brown
Oxide**   Uranium dioxide, UO_2. *See* Refined Uranium.

Decommissioning

All technical and administrative actions leading to the release of a facility from regulatory control. This usually includes preliminary characterization, preparation and licensing of the strategy and activities, clean-up, decontamination, and dismantling activities, segregation and packaging of radioactive and non-radioactive wastes, and the final radiological monitoring for release. See [175].

Drift   A mined-out region in an underground mine. These would usually be horizontal and/or parallel to the ore deposits. *See also* Stope.

Driftite   A "semi-gelatin dynamite," typically comprising 70% nitroglycerin mixed with ammonium nitrate and sodium nitrate, and a hydrocarbon like tar (to make it waterproof). *See also* Cilgel, Forcite.

Dygel   *See* Forcite.

EBR-I   *See* Experimental Breeder Reactor I.

Efflorescent Salts

Efflorescence refers to the spontaneous evaporation of water from a hydrated salt. When such salts effloresce, they may lose some or all of their water of hydration, and they generally take on a powdery appearance. When the efflorescing salts originate within a porous material like tailings, they tend to form a coating on the surface.

**Environmental
Remediation**   Activities aimed at reducing radiation exposure from existing or potential contamination of land areas. This usually includes actions aimed at the contamination itself (by reducing and/or confining the source) and/or at the pathways for human and environmental exposure. See [175].

EQC   *See* NSEQC.

Experimental Breeder
Reactor I   (EBR-I) The United States' first electric-power generating nuclear reactor, which was built in Idaho and started-up in December 1951.

Extract   *See* Raffinate.

Forcite   A "gelatin dynamite," comprising 30 to 80% nitroglycerin mixed with cellulose, sodium or potassium nitrate, and a hydrocarbon like tar (to make it waterproof). Cilgel and Dygel (both trademarks of CIL and ICI Canada) were other gelatin dynamite formulations. *See also* Cilgel, Driftite.

Geiger-Müller
Meter   One of the first commercial hand-held radiation detector/counters. The Geiger-Müller Meter uses an ionization-chamber detector of the same name, enabling it to detect alpha particles, beta particles, and gamma rays. Modern versions are still available today.

Green Salt Uranium tetrafluoride, UF_4. *See* Refined Uranium.

High Grading Selectively mining only the highest grades of material in an orebody.

Meteoric Water
                Water from precipitation, such as rain and snow. Meteoric water also includes surface water from lakes and rivers, which indirectly come from precipitation.

Mine Cage   *See* Cage.

Mine Dry   A mine-site building in which workers can change into and out of their working clothes and wearable equipment. Such facilities generally also include sinks, showers, toilets, lockers, and dirty-clothes baskets.

Mine Skip   *See* Skip.

Muck   Rock, including both ore and waste rock, that has been blasted from a mine face. Mucking refers to the gathering and transporting broken-up ore and waste rock in a mine.

Mucking   *See* Muck.

National Research Experimental
Reactor   (NRX) Canada's second nuclear reactor. It was built at Chalk River, Ontario and commenced operation in 1947. NRX was a 10 MW (later 42 MW) heavy-water-moderated research reactor. It was built and operated by the National Research Council until 1952 and thereafter by Atomic Energy of Canada Ltd. It was closed in 1993. *See also* National Research Universal Reactor and Zero-Energy Experimental Pile Reactor.

National Research Universal
Reactor   (NRU Reactor) Canada's third nuclear reactor. It was built at Chalk River, Ontario and commenced operation in 1957. NRU is a 135 MW heavy-water-moderated research reactor. As one of Canada's national science facilities, it is used to generate isotopes for medical diagnoses and/or treatments, to generate neutrons for the Canadian Neutron Beam Centre, and it is also used in the development of CANDU reactor fuels and materials. It is still in operation. *See also* National Research Experimental Reactor and Zero-Energy Experimental Pile Reactor.

NEW   Nuclear Energy Worker. *See* Atomic Energy Worker.

NPD Reactor   *See* Nuclear Power Demonstration Reactor.

NRU Reactor   *See* National Research Universal Reactor.

NRX Reactor   *See* National Research Experimental Reactor.

NSEQC   The Northern Saskatchewan Environmental Quality Committee, comprising representatives from the northern municipal and First Nation communities that are impacted by northern mining operations in Saskatchewan.

Nuclear Energy
Worker   (NEW) *See* Atomic Energy Worker.

Nuclear Power Demonstration

Reactor   (NPD Reactor) Canada's first electric-power generating nuclear reactor, which was built in Ontario and started-up in June 1962.

Orange Oxide This term normally refers to a blend of red and yellow iron oxides, however, in the uranium industry, the term has been used to refer to a form of uranium trioxide (γ-UO_3) that is bright orange in colour. The latter is more commonly referred to as "orange powder." *See also* Refined Uranium, Yellowcake.

Orange Powder

A fairly pure form of uranium trioxide, γ-UO_3. *See* Refined Uranium.

Polymer   A World War II-era code word for heavy water. *See* Atomic Code Words.

Radium Ore An older term used to refer to the uranium mineral pitchblende. In the 1920s and '30s, uranium minerals were of interest to prospectors as an indicator of radium potential.

Radon   Radon is a chemical element that occurs naturally as a decay product of radium, which in turn is a decay product of uranium. As a result, radium and radon tend to be found wherever there is uranium. Radon poses human health concerns, not so much from radon itself, but from the alpha particles emitted from its decay products: polonium-210 and polonium-214, which can be adsorbed onto fine solid particles and/or small water droplets in the air, then inhaled, and then trapped in the lungs. In this case, the alpha particles emitted can directly irradiate lung tissue, which can cause lung cancer [22,24].

Raffinate   In an industrial chemical separation process, where solvent extraction is used to remove components from a liquid, the phase containing the removed material is referred to as the extract, while the remaining liquid from which components have been removed is referred to as the raffinate. Depending upon the process, either phase may contain the desired product. In the case of the solvent extraction of uranium, the raffinate contains the unwanted residual components ("waste").

Raise A vertical, or nearly vertical, opening in an underground mine that leads from one level to another, and sometimes all the way to the surface.

Refined Uranium In the refining of uranium any of a number of compounds may be the final, or intermediate, product. In very early refineries, yellowcake was converted into a form of uranium trioxide called "black oxide" (which was about 95 percent U_3O_8). Later refineries produced a more pure form of UO_3, called "orange powder" (or "bright orange powder," or "orange oxide") due to its yellow-orange colour. In some refineries uranium trioxide was reduced to uranium dioxide, UO_2, called "brown oxide" (or "dark brown oxide") due to its dark brown colour, and then converted to uranium tetrafluoride, UF_4, called "green salt," due to its green colour. *See* Figure 6.3. *See also* Yellowcake.

Skip A bucket-like container, suspended from a hoist on steel wire rope and used to transport mined ore and waste rock up an underground mine shaft to the surface. Also called a Mine Skip. Skips were not used at the Nicholson Mine. Instead, tram cars were simply rolled into and out of, one or both of the cage compartments. *See also* Cage.

Saskatchewan Research Council (SRC) A research and technology organization incorporated as a Crown Corporation and owned by the Government of Saskatchewan. SRC conducts independent applied, research, development, demonstration, testing, and commercialization.

SRC *See* Saskatchewan Research Council.

Stope In underground mining a stope is the ore surface being mined and/or the open passageway space that is left behind after the ore has been mined. A near-horizontal such passageway is termed a drift. Stoping refers to the removal of the ore from this space and is practised when the surrounding rock is stable enough not to collapse after the ore has been mined out. *See also* Drift.

Tube Alloys   The code name for the secret atomic weapons development programs of the U.S., UK, and Canada, that were merged in 1943.

U-235   The specific isotope of uranium (U) that is involved in sustainable nuclear fission. U-235 is naturally present only in very low concentrations, less than one percent, in the main uranium isotope, which is U-238. The numbers refer to the relative atomic mass of the element – atoms of U-238 have three more neutrons in them than do atoms of U-235.

Vein   A mineral vein is a layer or sheet of crystallized minerals in a rock formation. Such veins would have been created by the precipitation of mineral components from a solution as it flowed through a natural fissure, or crack, in the rock.

Yellowcake   The final precipitated oxides of uranium that result from the milling of raw uranium ore using a leaching process. Although often referred to as U_3O_8, for older processes this is a bulk-average approximation. Yellowcake from Cold War-era milling operations was usually a mixture of UO_2 and UO_3 with minor amounts of uranyl hydroxide and uranyl sulphate. The "yellowcake" produced by some mills is (was) actually brown or black, rather than yellow in colour. *See also* Refined Uranium.

ZEEP Reactor *See* Zero-Energy Experimental Pile Reactor.

Zero-Energy Experimental Pile
Reactor   (ZEEP Reactor) Canada's first nuclear reactor and the world's first non-U.S. reactor. It was built at Chalk River, Ontario and commenced operation in 1945. ZEEP was a heavy-water-moderated reactor and was used to irradiate uranium to produce plutonium, and also to irradiate thorium to produce uranium-233. It was closed in 1970. *See also* National Research Experimental Reactor and National Research Universal Reactor.

11 APPENDICES

Appendix 1. General Longitudinal Section of the Mine (Lorado Uranium Mines Ltd., April 1, 1960. The arrows show the air flow directions. Note the four pyrite mining areas shown near the centre.

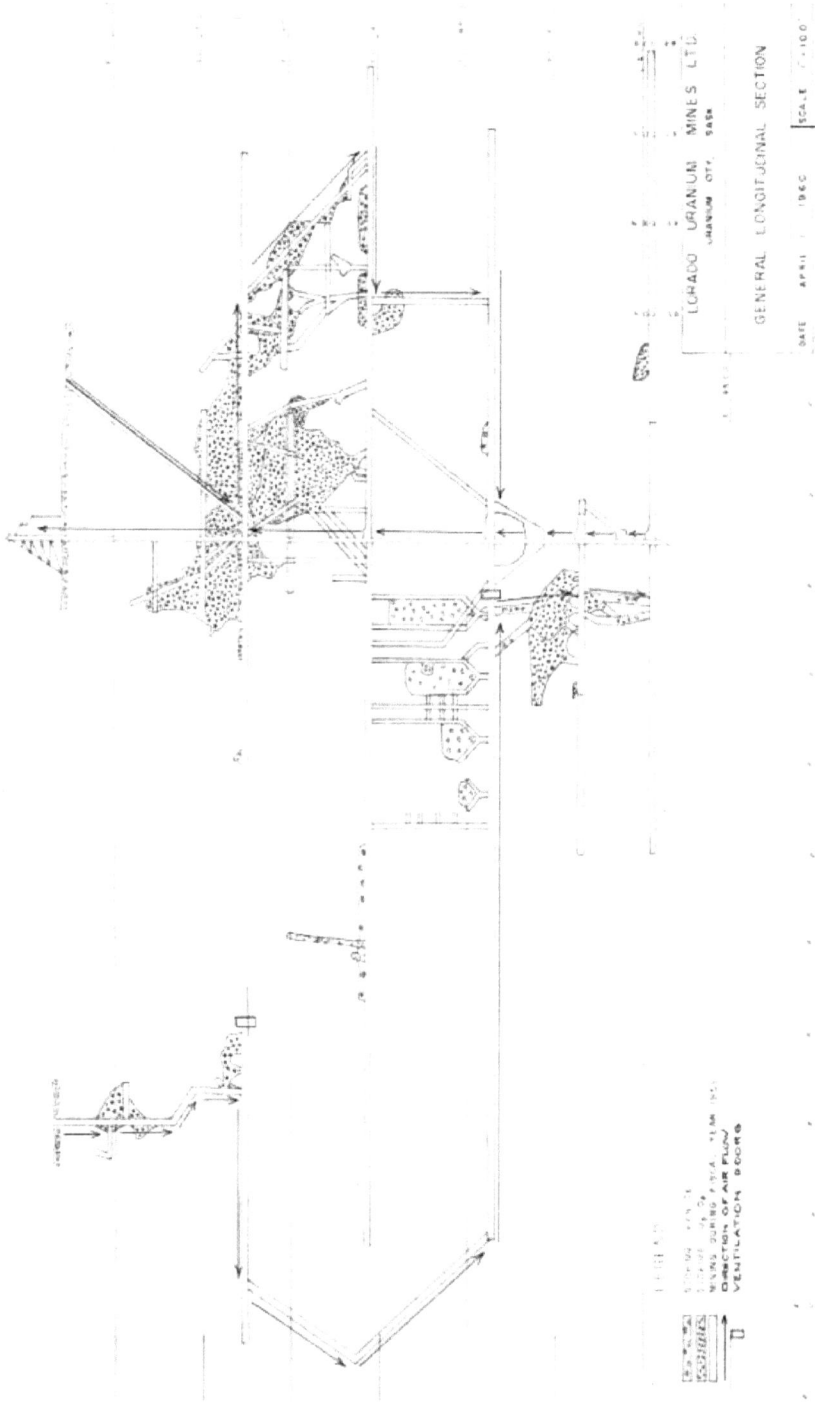

LORADO URANIUM MINES LTD
URANIUM CITY, SASK.

GENERAL LONGITUDINAL SECTION

DATE   APRIL   1960   SCALE 1"=100'

DIRECTION OF AIR FLOW
VENTILATION DOORS

Appendix 2. General Longitudinal Section of the Mine (Lorado Uranium Mines Ltd., April 13, 1960. This drawing shows where horizontal cut-and-fill stoping was used (on the second level), and where shrinkage stoping was used (on all levels).

Appendix 3. General Longitudinal Section of the Mine (Lorado Uranium Mines Ltd., January 10, 1959).

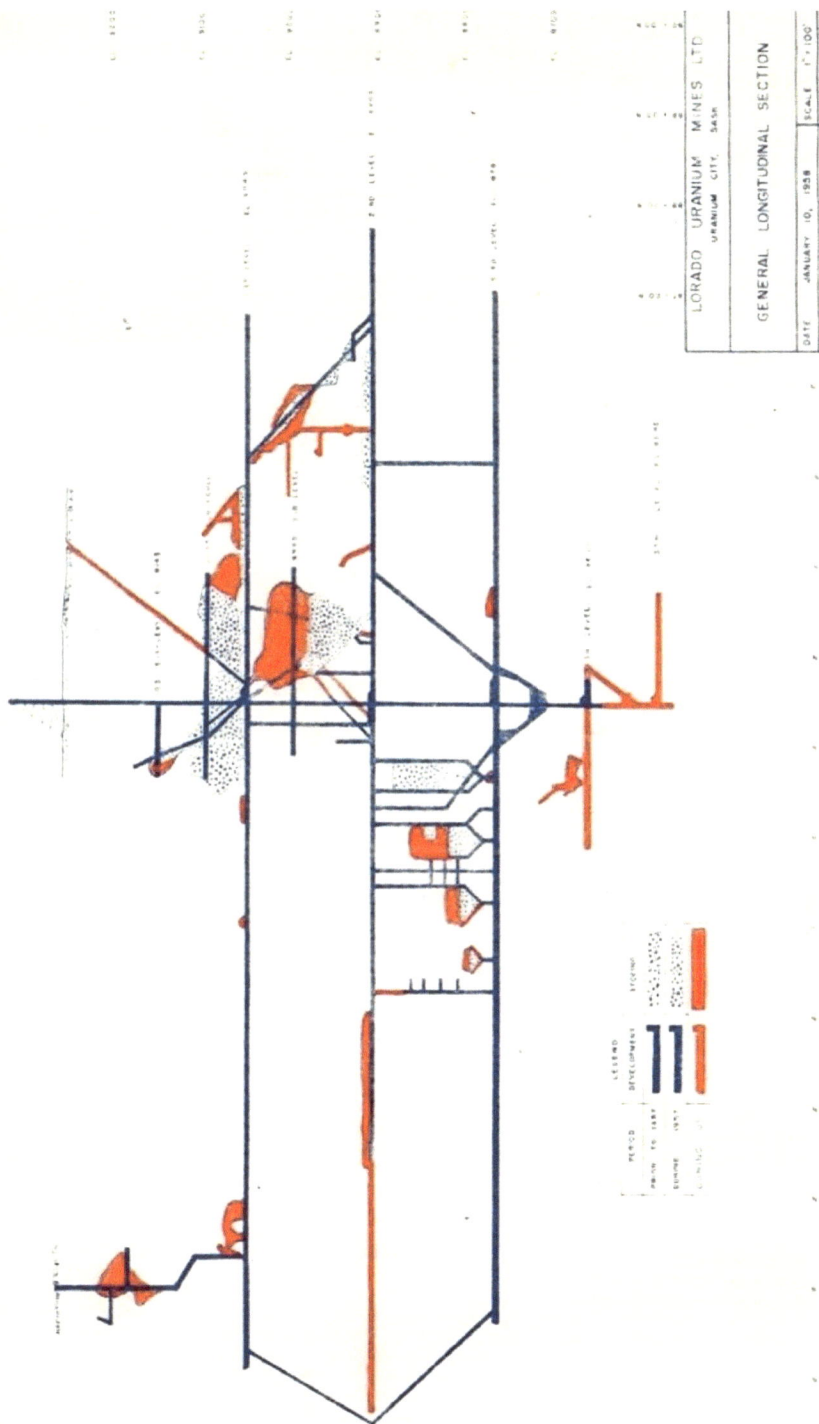

LORADO URANIUM MINES LTD.
URANIUM CITY, SASK.

GENERAL LONGITUDINAL SECTION

DATE JANUARY 10, 1958   SCALE 1"=100'

Appendix 4. Mine First Level Plan Showing Developments Up to 1958. (Lorado Uranium Mines Ltd., January 10, 1959).

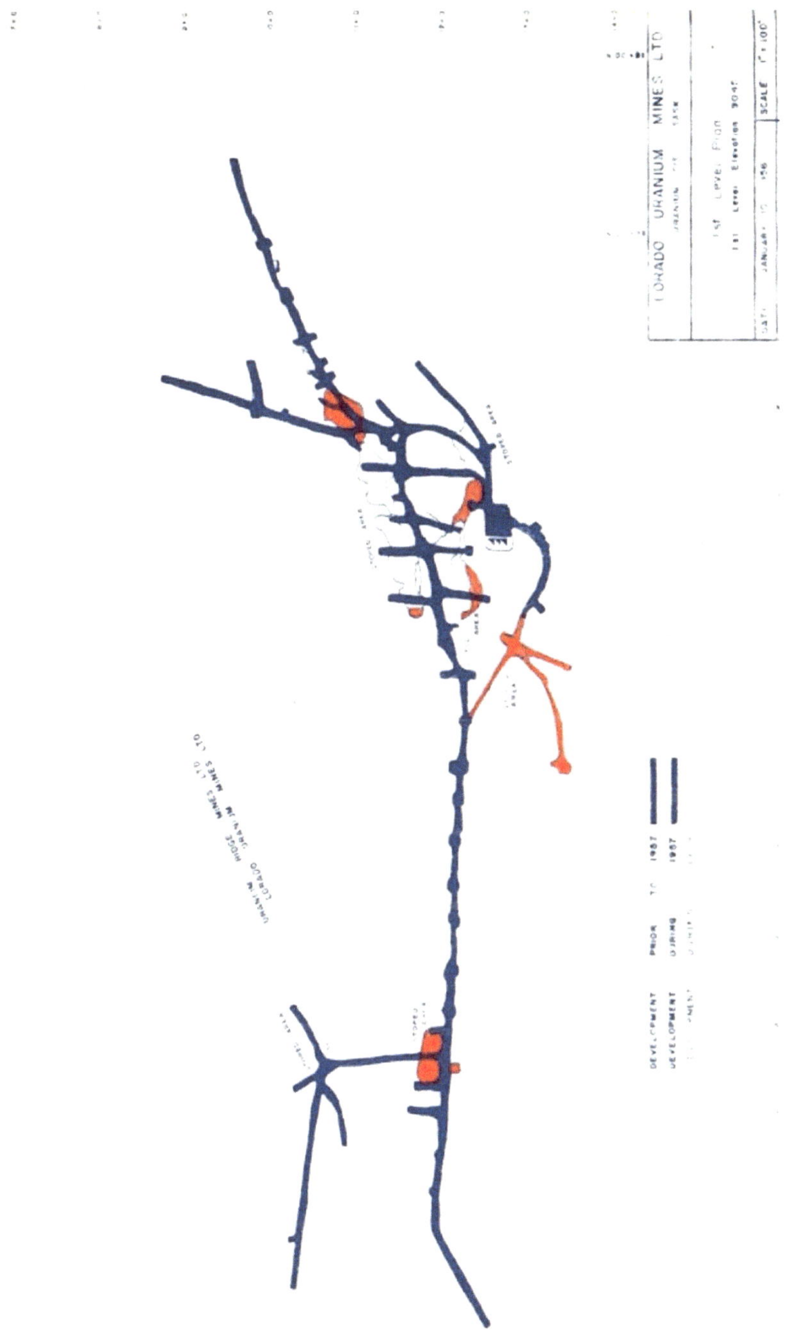
LORADO URANIUM MINES LTD
1st Level Plan
1st Level Elevation 9047
SCALE 1" = 100'
DEVELOPMENT PRIOR TO 1957
DEVELOPMENT DURING 1957

Appendix 5. Mine Second Level Plan Showing Developments Up to 1958. (Lorado Uranium Mines Ltd., January 10, 1959).

Appendix 6. Mine Third Level Plan Showing Developments Up to 1958. (Lorado Uranium Mines Ltd., January 10, 1959).

LORADO

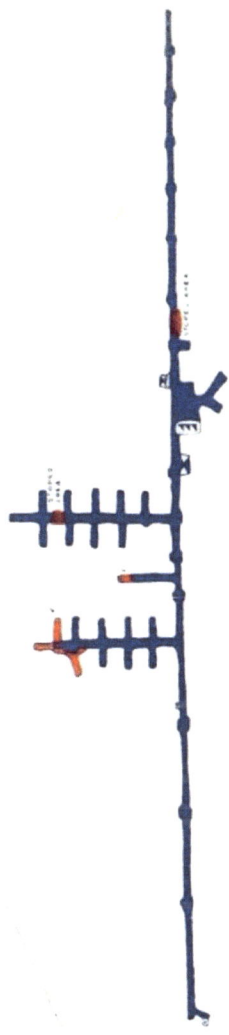

Appendix 7. Mine Fourth Level Plan Showing Developments Up to 1958. (Lorado Uranium Mines Ltd., January 10, 1959).

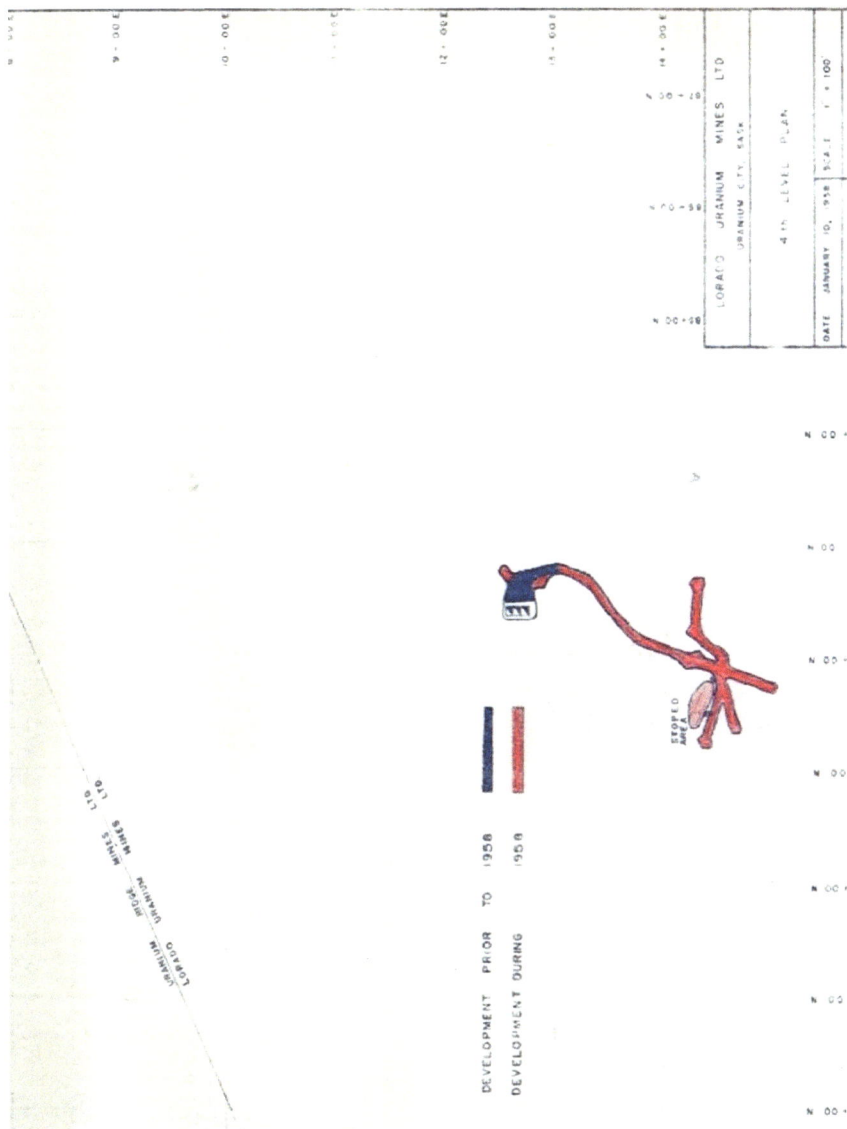

Appendix 8. Mine Fifth Level Plan Showing Developments Up to 1958. (Lorado Uranium Mines Ltd., January 10, 1959).

LORADO

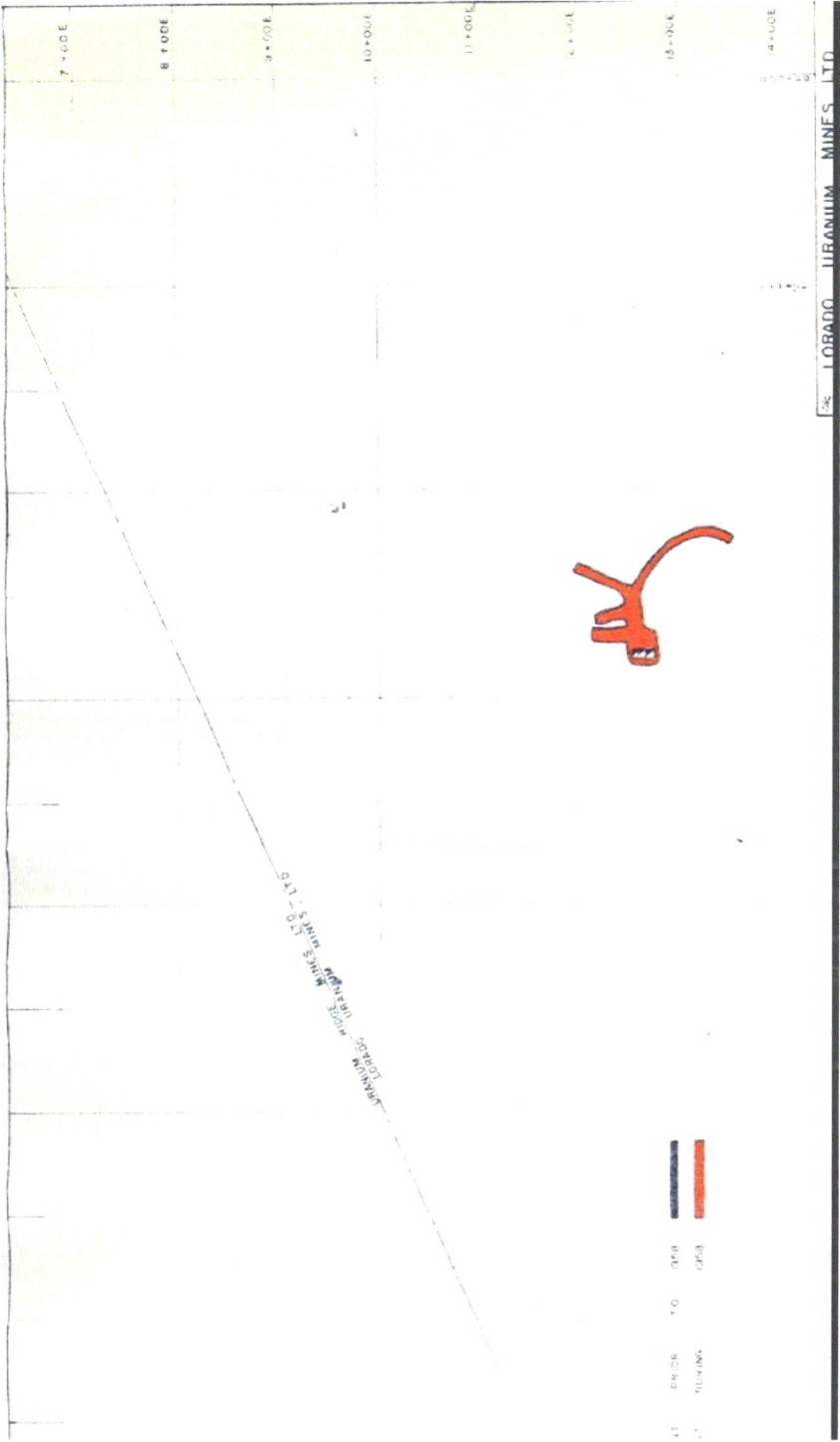

Appendix 9. Mine Composite Plan Showing Most of the Mined Drifts, at Various Working Levels. (Lorado Uranium Mines Ltd., April 11, 1959).

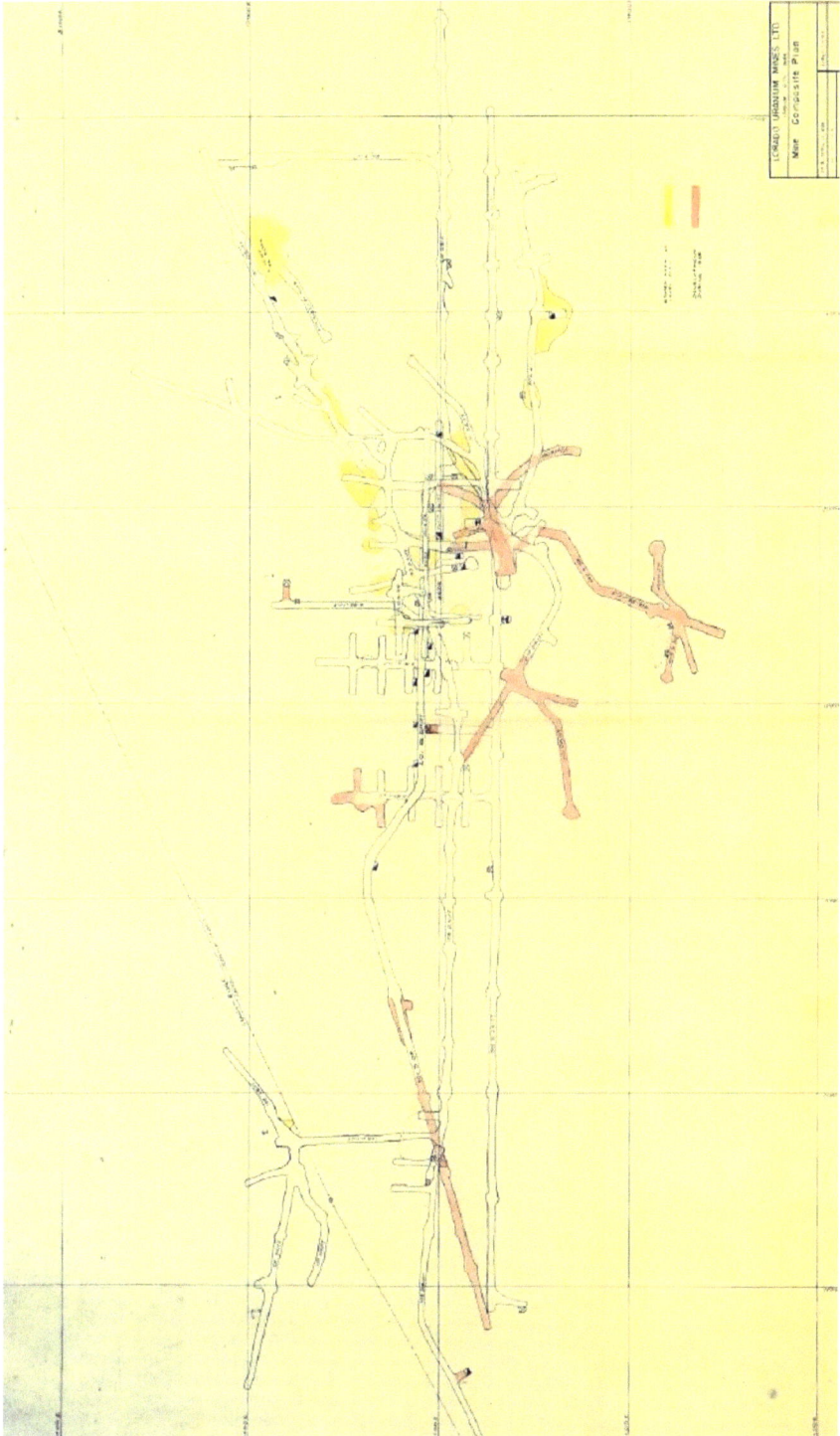

Appendix 10. Approximate Unit Conversions.

These unit conversions are approximate only:

Mass	Imperial pounds to kilograms	1 lb = 0.454 kg
	Imperial tons to metric tonnes	1 ton = 0.907 tonne
Distance	Imperial feet to metric metres	1 ft = 0.3048 m
Volume	U.S. gallons to metric litres	1 US gal = 3.785 l
	Imperial gallons to metric litres	1 Imp gal = 4.546 l

For uranium, 1 tonne of uranium metal in U_3O_8 = 1.1792 tonne as U_3O_8.

12 SUMMARY

Lorado,
A Saskatchewan Cold War Uranium Mine and Custom Mill

The Lorado deposit was discovered in 1950 in a remote location in northern Saskatchewan, and just north of Lake Athabasca. Subsequent exploration and development led to a mine, mill, and associated campsites being built, all of which were fully operational by 1957. The Lorado mill was unique, having been designed to process ores from smaller, neighbouring mines that would otherwise not have succeeded. This made Lorado the third largest producer of uranium (yellowcake) concentrate in Saskatchewan and one of the top five in Canada during the Cold War era. By 1960 the markets for uranium had crashed and operations were closed but, having produced about 1,210 tonnes of uranium concentrate, Lorado had played a significant role in helping Canada become one of the largest uranium producers in the world. Beyond uranium, the Lorado mine produced about 500,000 m^3 of highly acidic tailings, which entered nearby Nero Lake virtually destroying it. Following closure in 1960, the Lorado site stood abandoned for the next twenty years, until the site owners cleaned up most of the mine infrastructure in 1982, and the mill buildings in 1990. Another sixteen years would pass before the government of Saskatchewan stepped in and contracted the management of the rest of the remediation to the Saskatchewan Research Council (SRC). At the time of writing this book essentially all of the Lorado sites' remediation had been completed, with active monitoring in progress subsequent to ultimately releasing them into a long-term management and monitoring program.

Print ISBN: 978-0-9958081-6-4
ePub ISBN: 978-0-9958081-7-1

13 ABOUT THE AUTHORS

Dr. Laurier Schramm has over 35 years of R&D experience spanning each of the industry, not-for-profit, university, and government sectors. He is currently President and CEO of the Saskatchewan Research Council (SRC). His interests include technological innovation, management and leadership, colloid & interface science, and nanotechnology. He holds 17 patents and has published 16 books and over 400 other publications and proprietary reports. He has served on many expert advisory panels and Boards, is a co-founder of Innoventures Canada Inc. (I-CAN), and co-founder of Canada's Innovation School™. He has received national scientific and engineering awards for his work and is a Fellow of the Chemical Institute of Canada and an honourary Member of the Engineering Institute of Canada.

Patty Ogilvie-Evans was born and raised in Saskatchewan, and her interest in earth processes led her to pursue a career in Geological Sciences. She graduated from the University of Saskatchewan in 2006 and has been working as a Geologist in Saskatchewan for the past 12 years. She has a mining background with experience in gold and uranium underground and open-pit mining, diamond exploration and several years of uranium exploration. She is currently a part of Saskatchewan Research Council's (SRC) Environmental Remediation team, working with abandoned legacy sites in Northern Saskatchewan. The history of the abandoned sites is of particular interest to Patty and she greatly enjoys the challenges of finding, locating and recreating abandoned mines to provide a better understanding to assist in the ultimate remediation of these sites.

Ian Wilson is the Remediation Manager, in the Environment Division, at the Saskatchewan Research Council. Ian has an MBA from London School of Business and Finance and a B.Sc. in Environmental Science from Royal Roads University. He has previously served as Regional Manager at Quantum Murray LP, and Project Scientist with both SNC-Lavalin Environment and O'Connor Associates (Parsons). Ian has more than 17 years of environmental remediation experience and has successfully managed more than 200 assessment, remediation and site decommissioning projects around the world. Ian also provides technical and policy development advice for various international organizations. Areas of expertise include remediation design, remediation project management, cost estimation, structural demolition, mine closure, stakeholder engagement, and waste management.

14 REFERENCES

1. Eldorado, *Uranium in Canada*, Eldorado Mining and Refining Ltd.: Ottawa, 1964.
2. Penrose, R.A.F., *Econ. Geol.*, **1915**, *10*, 161-171.
3. Hahn, O., "From the Natural Transmutations of Uranium to its Artificial Fission," In *Nobel Lectures, Chemistry 1942-1962*, Elsevier, Amsterdam, 1964, pp. 51-66 (Hahn's Nobel Prize Lecture given on 13 Dec. 1946).
4. Ringholz, R.C., *Uranium Frenzy, Boom and Bust on the Colorado Plateau*, Norton: NY, 1989.
5. Amundson, M., *"Yellowcake Towns: Uranium Mining Communities in the American West,"* University Press of Colorado, Boulder, 2004.
6. Davidson, C.F., *The New Scientist*, **1957**, *(Feb. 21)* 9-11.
7. Taft, R.B., *Radium Lost and Found*, Furlong: Charleston, 1938.
8. Griffith, J.W., *The Uranium Industry - Its History, Technology and Prospects*, Mineral Report 12, Dept. of Energy, Mines and Resources: Ottawa, 1967.
9. Guidry, M., *The Guedry-Labine Family and the World's First Atomic Bomb*, accessed December 2013, http://freepages.genealogy.rootsweb.ancestry.com/~guedrylabinefamily /guedrylabineatomicbomb.
10. Bothwell, R., *Eldorado, Canada's National Uranium Company*, University of Toronto Press: Toronto, 1984.
11. Globe and Mail, "Gilbert LaBine: His Tools Were Pick, Paddle and .30-.30," *The Globe and Mail*, **1957**, *July 20*, p. 35.
12. Saskatchewan Department of Mineral Resources, *Inventory and Outlook of Saskatchewan's Mineral Resources*, Report No. 83, Dept. Mineral Resources: Regina, SK, Nov. 1966, 52 pp.
13. Natural Resources Canada, *Atlas of Canada*, 6th Ed., Natural Resources Canada: Ottawa, 2009.
14. Alcock, F.J., "Geology of Lake Athabaska Region, Saskatchewan," Memoir 196, Geological Survey of Canada, Ottawa, 1936.

15. CIM, *The Beaverlodge Uranium District,* Beaverlodge Branch, Canadian Institute of Mining & Metallurgy, Edmonton, Sept. 1957, 57 pp.
16. Saskatchewan Geological Survey, "Geology, and Mineral and Petroleum Resources of Saskatchewan 2003," Saskatchewan Industry and Resources: Regina, Misc. Report 2003-7, 2003.
17. Beck, L.S., "Uranium Deposits of the Athabasca Region," Report 126, Geological Survey, Saskatchewan Mineral Resources: Regina, 1969.
18. La Bine, D.G., *Gilbert A. LaBine 1890 – 1977,* accessed December 2013, http://www.labine.com/gilbert_a_labine, 2004.
19. Hahn, O.; Strassmann, F. "Concerning the Existence of Alkaline Earth Metals Resulting from Neutron Irradiation of Uranium" *Naturwiss.,* **1939**, *27,* 11-15. Translation in *Am. J. Phys.,* **1964**, *January,* 9-15.
20. Meitner, L.; Frisch, O. R. "Disintegration of Uranium by Neutrons: a New Type of Nuclear Reaction," *Nature,* **1939**, *143 (3615),* 239–240.
21. Peierls, R. "O. R. Frisch, 1904-1979," *Nature,* **1980**, *284 (13 March),* 196–197.
22. Zoellner, T., *Uranium,* Penguin Books: London, 2009.
23. Rutherford, E., *Radio-Activity,* Cambridge University Press: Cambridge, 1904.
24. Sims, G.H.E., *A History of the Atomic Energy Control Board,* Canadian Government Printing Centre: Ottawa, 1980.
25. Dominion Bureau of Statistics, "Chronological Record of Canadian Mining Events from 1604 to 1943 and Historical Tables of the Mineral Production of Canada," Department of Trade and Commerce, Edmond Cloutier Printer: Ottawa, ON, 1945.
26. CIM Staff, "The Eldorado Enterprise," *Trans. Can. Inst. Min. Met.,* **1946**, *49,* 423-438.
27. World Nuclear Association, *Brief History of Uranium Mining in Canada,* Appendix 1, World Nuclear Association: London, accessed January 2013, http://www.world-nuclear.org/info/Country-Profiles/Countries-A-F/Appendices/Uranium-in-Canada-Appendix-1--Brief-History-of-Uranium-Mining-in-Canada/.
28. Piper, L., *Environment and History,* **2007**, *13,* 155-186.
29. "Early Instrumentation - 1920's," *National Radiation Instrument Catalog 1920 – 1960,* 2007, http://national-radiation-instrument-catalog.com/new_page_144.htm.
30. Taft, R.B., "Radium Hounds," *Scientific American,* **1939**, *160(1),* 8-47.
31. LaBine, G.A., "Submission to Royal Commission on Canada's Economic Prospects," Government of Canada, Ottawa, 8 March 1956.
32. Hunter, W.D.G., "The Development of the Canadian Uranium Industry: An Experiment in Public Enterprise," *Can. J. Econ. Pol. Sci.,* **1962**, *28(3),* 329-352.
33. Ross, M.; Hovdebo, D.G., "Uranium Mine Reclamation - A Myriad of Extremes Politics, Perceptions and Long-Lived Radionuclides," in Proc.19th Annual British Columbia Mine Reclamation Symposium, Dawson Creek, BC, pp. 188-196 (1995).

34. Athabasca Interim Advisory Panel, "Athabasca Land Use Plan: Stage One," Saskatchewan Environment: Regina, March 2006.

35. Lang, A.H., "History of Uranium Discoveries, Lake Athabasca," *Can. Min. J.*, **1953**, *74(6)*, 69-76.

36. Kneen, J., "Uranium Mining in Canada – Past and Present," Presented to *Indigenous World Uranium Summit*, Nov. 30-Dec. 1, 2006, Window Rock, Arizona. Accessed at: http://www.miningwatch.ca/sites/www.miningwatch.ca/files/Uranium _Canada_0/

37. Schramm, L.L.; Ogilvie-Evans, P., *The Nicholson Mine. Saskatchewan's First Cold War Uranium Mine*, Saskatchewan Research Council, Saskatoon, 2018.

38. Tilman, A., "On the Yellowcake Trail," Parts 1-4, *Watershed Sentinel*, **2009**, *June-July*, 18-22; **2009**, *Sept.-Oct.*, 28-31; **2009**, *Nov.-Dec.*, 28-31; and **2009**, *Mar.-Apr.*, 28-31.

39. Belanger, D.; Hallett, F.; Dusseault, C., *The History of Uranium City*, Self-published. Available through several public libraries including the La Ronge Public Library and the Saskatoon Public Library, 1975, 19 pp.

40. MiningWatch Canada, Elliot Lake Uranium Mines, MiningWatch Canada: Ottawa, 2012, http://www.miningwatch.ca/elliot-lake-uranium-mines.

41. Hutton, E., "The Atom Bomb That Saves Lives," *Maclean's Magazine*, **1952**, *65(4)* February 15, p.7.

42. Fedoruk, S., "The Growth of Nuclear Medicine," *50 Years of Nuclear Fission in Review*, Canadian Nuclear Society: Ottawa, 1989, http://media.cns-snc.ca/history/fifty_years/fedoruk.html.

43. Idaho National Laboratory, "Experimental Breeder Reactor - I (EBR-I)," Brochure 07-GA50535_02, Idaho National Laboratory: Idaho, 2007.

44. Fawcett, R., *Nuclear pursuits: The scientific biography of Wilfred Bennett Lewis*, McGill-Queen's University Press: Montreal, 1994.

45. Grade 10 Class Candu High School, *The History of Uranium City and District,* Lakeland Press, La Ronge, SK, 1982, 63 pp.

46. McBain, L., "Uranium City," Encyclopedia of Saskatchewan, University of Regina: Regina, 2006, http://esask.uregina.ca/entry/uranium_city.html.

47. Nichiporuk, A., "What Does the Future Hold for Uranium City?" *CIM Magazine*, **2007**, (November), 2 pp.

48. Keeling, A., "Born in an atomic test tube. Landscapes of cyclonic development at Uranium City: Saskatchewan," *The Canadian Geographer,* **2010**, *54(2)*, 228-252.

49. Northern Miner, "Saskatchewan Uranium Shows Attract Monied Interest," *The Northern Miner*, 1952, Jan. 10, p. 1.

50. "The Uranium Rush - 1949," *National Radiation Instrument Catalog 1920 – 1960*, 2007, http://national-radiation-instrument-catalog.com/new_page_144.htm.

51. US Atomic Energy Commission, *Prospecting For Uranium*, US Government Printing Office: Washington, 1949.

52. Wright, R.J., *Prospecting with a Counter*, U.S. Atomic Energy Commission: Washington, 1954.
53. Northern Miner, "Uranium – Canada Maintains Place in Frantic World Production Race," *The Northern Miner*, 1952, Nov. 27, p.58.
54. Joubin, F.R.; James, D.H., "Canada's Uranium Future," *Precambrian*, **1956**, *29(5)*, 13-14.
55. Maclean's, "Uranium City Here We Come," *Reader's Digest Magazine*, **1954**, *64(384)*, April, 59-64.
56. Richardson, B.T., "The Hottest Square Mile in the World," Maclean's Magazine, **1951**, *64(20) Oct. 15, p. 14.*
57. Life, "Uranium Rush is On in Athabaska," *Life Magazine*, **1952**, *33(7)*, Aug. 18, pp. 15-19.
58. Stapleton, B., "Canada's Great Uranium Rush," *Collier's Magazine*, **1953**, *October 2*, pp. 32-41.
59. Advocate, "Atom Age Mining Rush Begins in N. Canada," *Advocate (Burnie, Tasmania)*, **1952**, *August 5*, p. 3.
60. Courier-Mail, "Began at dawn, Uranium rush in Canada," *The Courier-Mail (Brisbane, Queensland)*, **1952**, *August 5*, p. 1.
61. Sydney Morning Herald, "Canada's First Uranium Rush," *The Sydney Morning Herald (New South Wales)*, **1952**, *August 5*, p. 3.
62. Mercury, "Uranium rush in Canada," *The Mercury (Hobart, Tasmania)*, **1952**, *August 5*, p. 3.
63. TMC, *"The Birth of a Great Uranium Area,"* Documentary Film, Technical Mine Consultants (TMC, Toronto) and Canadian Television Film Production, 1953.
64. Northern Miner, "Many Companies Active in Sask.," *The Northern Miner*, 1954, Sept. 16, p.2.
65. ITN, *"The Road to Uranium,"* Documentary Film, Independent Television News (ITN), London, U.K., 16 October 1957.
66. British Columbia Geological Survey, "MINFILE Mineral Inventory," MINFILE Record Summary, MINFILE No 082M 021, British Columbia Ministry of Energy and Mines: Victoria, BC, 2013, http://minfile.gov.bc.ca/Summary.aspx?minfilno=082M++021.
67. Piper, L., *The Industrial Transformation of Subarctic Canada*, UBC Press: Vancouver, 2009.
68. Gunnar Mines Ltd., *The Gunnar Story*, Gunnar Mines Ltd., Toronto, Sept. 1957.
69. Delaney, G., "Uranium in Saskatchewan, Canada," Proc. South Australian Resources and Energy Investment Conference (SAREIC 2009), Unlocking South Australia's Mineral Wealth, 6 May 2009, http://www.pir.sa.gov.au/__data/assets/pdf_file/0006/104559/Gary_Delaney.pdf.
70. Schramm, L.L., *Gunnar Uranium Mine: Canada's Cold War Ghost Town*, Saskatchewan Research Council, Saskatoon, 2016.
71. Gunnar Mines, "25th Annual Report. For the Year 1958," Gunnar Mines Ltd., Toronto, 23 March 1959.

72. SRC, *12th Annual Report of the Saskatchewan Research Council 1958*, Saskatchewan Research Council, Regina, 1959.

73. Cipriani, A.J., "Radiation Hazards in Uranium Mines," *Can. Mining J.*, **1955**, *76(9)*, 79-80.

74. Schramm, L.L., *Research and Development on the Prairies. A History of the Saskatchewan Research Council*, Saskatchewan Research Council, Saskatoon, 2016.

75. Muldoon, J.A., "Policy Networks, Policy Change and Causal Factors, A Uranium Mining Case Study," Ph.D. Thesis, University of Regina, Regina, SK, March 31, 2015.

76. World Nuclear Association, *"Uranium in Canada,"* World Nuclear Association: London, U.K., January 2016, http://www.world-nuclear.org/info/country-profiles/countries-a-f/canada--uranium.

77. Natural Resources Canada, "About Uranium," Natural Resources Canada, Ottawa, January 2016, http://www.nrcan.gc.ca/energy/uranium-nuclear/7695.

78. Brean, H., "Penny Stocks for Dollar Profits," *Life Magazine*, **1953**, *34(23)*, 8 June, pp. 143-155.

79. Watters, R.; McKee, P.; Lush, D., "An Evaluation of Potential Environmental and Public Safety Impacts of Gunnar and Lorado Facilities in Northern Saskatchewan. Vol. 1. Summary of Existing Baseline Data," Report for Saskatchewan Environment by Beak Consultants Ltd., Sept. 1989.

80. Lang, A.H.; Griffith, J.W.; Steacy, H.R., "Canadian Deposits of Uranium and Thorium," Economic Geology Series No. 16, 2nd Ed., Geological Survey of Canada, Ottawa, 1962.

81. Robinson, S.C., Mineralogy of Uranium Deposits, Goldfields, Saskatchewan," Bulletin 31, Geological Survey of Canada, Ottawa, 1955.

82. Lorado, "Annual Report for the Year Ending April 30, 1956," Lorado Uranium Mines Ltd., Toronto, 1 September 1956.

83. Northern Miner, "Lorado Uranium Mines to Drill Next Month," *The Northern Miner*, 1952, May 22, p. 19.

84. Northern Miner, "Report Radioactivity in Lorado Drilling," *The Northern Miner*, 1952, July 3, p. 14.

85. Northern Miner, "Lorado Uranium Mines Enlarges Holdings," *The Northern Miner*, 1952, Aug. 21, p. 7.

86. Northern Miner, "Two More Holes at Lorado Ur," *The Northern Miner*, 1953, Aug. 6, pp. 17-18.

87. Northern Miner, "Radioactivity in Lorado Hole," *The Northern Miner*, 1953, Aug. 20, pp. 17-18.

88. Northern Miner, "Lorado Uranium Extending Zone," *The Northern Miner*, 1953, Dec. 17, p. 3.

89. Northern Miner, "Lorado Uranium is Encouraged," *The Northern Miner*, 1953, Aug. 13, p. 3.

90. Northern Miner, "Lorado Uranium Values Continue," *The Northern Miner*, 1954, Oct. 7, p. 7.

91. Oliver, K.S., "Lorado Considers Mill Construction," *Western Miner*, **1955**, *28(10)*, 77.

92. Northern Miner, "Lorado Going Underground," *The Northern Miner*, 1954, May 27, p. 1.

93. Northern Miner, "Picture Grows at Lorado," *The Northern Miner*, 1954, Nov. 25, p. 114.

94. Northern Miner, "Big Ore Width at Lorado," *The Northern Miner*, 1954, Dec. 9, p. 16.

95. Northern Miner, "Lorado Drifting Lengthens Ore," *The Northern Miner*, 1954, Dec. 16, p. 3.

96. Northern Miner, "Lorado Mill Decision by End of Year," *The Northern Miner*, 1955, Aug. 18, p. 16.

97. Northern Miner, "Lorado May Build New Custom Mill; Funds Arranged," *The Northern Miner*, 1955, Oct. 6, pp. 1,8.

98. Northern Miner, "Lorado Seeking Sales Agreement for Output From Custom Mill," *The Northern Miner*, 1955, Oct. 27, p. 1.

99. Northern Miner, "Lorado Boosts Capital," *The Northern Miner*, 1956, Apr. 19, p. 7.

100. Abelson, A., "After the Fall-Out. Uranium is Becoming a Big, Respectable Business," Barron's, 15 October 1956, pp. 36, 42.

101. Lorado, "Annual Report for the Year Ending April 30, 1958," Lorado Uranium Mines Ltd., Toronto, 30 September 1958.

102. Wall Street Journal, "Lorado Uranium Mines Issue," *Wall Street Journal*, **1956**, 27 August, p. 11.

103. Mamen, C., "Uranium Mining Methods," *Can. Mining J.*, **1956**, *77(6)*, 89-117, 156-157.

104. Eldorado, 1948 Annual Report, Eldorado Mining and Refining Ltd., Ottawa, 1949.

105. Wright, J., "Saskatchewan's North," *Can. Geog. J.*, **1952**, *45(1)*, 14-33.

106. Northern Miner, "Record Tonnage Moves Down North to Uranium Mines," *The Northern Miner*, 1956, Oct. 4, pp.17,20.

107. Gunnar Mining Ltd., "The Gunnar Story," *Can. Mining J.*, **1963**, *7*, 53-119.

108. Quiring, D.M., *CCF Colonialism in Northern Saskatchewan: Battling Parish Priests, Bootleggers, and Fur Sharks*, UBC Press, Vancouver, 2004.

109. Eldorado, 1949 Annual Report, Eldorado Mining and Refining Ltd., Ottawa, 1950.

110. Eldorado, "Annual Report 1960," Eldorado Mining and Refining Ltd., Ottawa, 1961.

111. Lorado, "Annual Report for the Year Ending April 30, 1957," Lorado Uranium Mines Ltd., Toronto, 9 October 1957.

112. Northern Miner, "See Lorado's New Custom Mill Ushering New Beaverlodge Era," *The Northern Miner*, 1957, Aug. 15, pp. 1,9.

113. Lorado, "Annual Report for the Year Ending April 30, 1959," Lorado Uranium Mines Ltd., Toronto, 24 September 1959.

114. Lorado, "Annual Report for the Year Ending April 30, 1960" Lorado Uranium Mines Ltd., Toronto, 6 October 1960.

115. Lorado, "Annual Report for the Year Ending April 30, 1961" Lorado Uranium Mines Ltd., Toronto, 3 October 1961.

116. Tremblay, L.P., "Geology of the Beaverlodge Mining Area, Saskatchewan," Memoir 367, Geological Survey of Canada: Ottawa, 1978.

117. Lorado, "Annual Report for the Year Ending April 30, 1962" Lorado Uranium Mines Ltd., Toronto, 9 October 1962.

118. Oulahan, R.; Lambert, W., "The Scandal in the Bahamas," *Life Magazine*, **1967**, *Feb. 3*, pp. 58, 60 – 74
(see http://www.jabezcorner.com/grand_bahama/oulaha3.htm).

119. Block, A.A., *Masters of Paradise: Organized Crime and the Internal Revenue Service in the Bahamas*, Transaction Publishers, Piscataway, N.J., 1991.

120. Lorado, "Annual Report for the Year Ending April 30, 1963" Lorado Uranium Mines Ltd., Toronto, 17 October 1963.

121. Lorado, "Annual Report for the Year Ending April 30, 1964" Lorado Uranium Mines Ltd., Toronto, 7 October 1964.

122. Lorado, "Annual Report for the Year Ending April 30, 1965" Lorado Uranium Mines Ltd., Toronto, 17 December 1965.

123. Lorado, "Annual Report for the Year Ending April 30, 1966" Lorado Uranium Mines Ltd., Toronto, 11 October 1966.

124. Lorado, "Annual Report for the Year Ending April 30, 1967" Lorado Uranium Mines Ltd., Toronto, 11 October 1967.

125. Mogul, "Annual Report 1968" International Mogul Mines Ltd., Toronto, 4 June 1969.

126. Mogul, "Annual Report December 31, 1980" International Mogul Mines Ltd., Toronto, 11 May 1981.

127. Robinson, S. C., "Mineralogy of the Goldfields District, Saskatchewan, Interim Account" Paper 50-16, Geological Survey of Canada, Ottawa, 1950.

128. Christie, A.M., "Gold Fields and Martin Lake Map-Areas, Saskatchewan," Paper 49-17, Geological Survey of Canada, Ottawa, 1949.

129. Lang, A.H., "Canadian Deposits of Uranium and Thorium, Interim Account" Economic Geology Series No. 16, 1st Ed., Geological Survey of Canada, Ottawa, 1952.

130. Jensen, K.A., "Technical Report on the Uranium City Properties for Uranium City Resources Inc. in the Uranium City Area NTS Map Sheets 74N-07, 74N-08, 74N-09 and 74N-10 and the Tazin Lake Area NTS Map Sheets 74N-14 and 74N-15 Northern Mining District Saskatchewan, Canada," Red Rock Energy Inc., Calgary, 2005, http://www.redrockenergy.ca/Technical_Report_-_Uranium_City_Properties.pdf.

131. MacDonald, B.C.; Kermeen, J.S., "The Geology of Beaverlodge," *Can. Mining J.*, **1956**, *77(6)*, 80-83, 156.

132. MacDonald, B.C., "Economic Geology of the Beaverlodge Uranium Area, Saskatchewan," *CIM Bull.*, **1956**, *49(5)*, 377-379.
133. Rich, R.A.; Holland, H.D.; Petersen, U., *Hydrothermal Uranium Deposits*, Elsevier, Amsterdam, 1977.
134. Tremblay, L.P., "Ore Deposits Around Uranium City," In *Structural Geology of Canadian Ore Deposits*, Symposium Proceedings, 6th Commonwealth Mining & Metallurgical Congress, CIM, Ottawa, 1957, pp. 211-220.
135. Lang, A.H., "Uranium Orebodies. How Can More be Found in Canada?" *Can. Min. J.*, **1952**, 73(6), 57-65.
136. KHS Environmental Management Group, "An Assessment of Abandoned Mines in Northern Saskatchewan," Report for Saskatchewan Environment, Regina, March 2001.
137. Joliffe, A.W., "The Gunnar 'A' orebody," *CIM Trans.*, *1956*, 59, 181-185.
138. Editorial, "People in the Uranium Industry," *Can. Mining J.*, **1956**, *77(6)*, 154-155.
139. Dyment, A.E., "Use of Explosives in Canadian Mining Operations," In *Mining in Canada*, Symposium Proceedings, 6th Commonwealth Mining & Metallurgical Congress, CIM, Ottawa, 1957, pp. 317-333.
140. Feasby, D.G., "Environmental Restoration of Uranium Mines in Canada: Progress Over 52 Years," *Proc. Workshop on Planning for Environmental Restoration of Uranium Mining and Milling Sites in Central and Eastern Europe*, International Atomic Energy Agency, Vienna, Nov. 1997, pp. 35-48.
141. Saskatchewan Research Council, KHS Environmental Management Group, and CanNorth Environmental Services, "Gunnar Site Characterization and Remedial Options Review," SRC Publication No. 11882-1C04, Saskatoon, January 2005.
142. Kennedy, E.G., *Oh Well, What the Hell!*, an autobiography, produced by BookBound Publishing, Macksville, New South Wales, 2002, but apparently never formally published.
143. Gunnar Mining Ltd., "The Gunnar Story," *Can. Mining J.*, **1963**, 7, 53-119.
144. UER.CA, "Uranium City," Urban Exploration Resource, UER.CA, 2014, http://www.uer.ca/locations/show.asp?locid=23608.
145. Woodward, J.R., "Lorado's Custom Mill," *Can. Mining J.*, **1956**, *77(6)*, 144-145.
146. Harper, C.T., "Northern Iron Ore Exploration," in "Summary of Investigations 1976," Report, Saskatchewan Geological Survey, Regina, 1976, pp. 104-114.
147. Grant, N.A., "General Mining Conditions at Eldorado Beaverlodge," *Trans. Can. Inst. Min. Met.*, **1953**, *56*, 260-263.
148. Butler, R.D., "Carbonate Leaching of Uranium Ores. A Review," Paper III, Proc. AAEC Symp. on Uranium Processing, Australian Atomic Energy Commission, Lucas Heights, Australia, 20-21 July 1972 pp. III-1 to III-3.
149. Smith, H.W.; Bull, W.R., "Uranium Ore Research," *Can. Mining J.*, **1956**, *77(6)*, 127-130, 159-160.

150. Thunaes, A., "Recovery of Uranium from Canadian Ores," *CIM Bull.*, **1954**, *47(3)*, 128-131.

151. Thunaes, A., "Review of Uranium Treatment Plants," In *The Milling of Canadian Ores*, Symposium Proceedings, 6th Commonwealth Mining & Metallurgical Congress, CIM, Ottawa, 1957, pp. 331-339.

152. Proulx, M., "The Uranium Mining Industry of the Bancroft Area: An Environmental History and Heritage Assessment," M.A. Thesis, Canadian Heritage and Development Studies Program, Trent University, Peterborough, May 1997.

153. Thunaes, A., "Uranium Recovery Plants," *Can. Mining J.*, **1956**, *77(6)*, 123-126,159.

154. Clegg, J.W.; Foley, D.D. (Eds.), *Uranium Ore Processing*, Addison-Wesley, Reading, 1958.

155. Woodward, J.R., "Lorado Uranium Mines Limited," In *The Milling of Canadian Ores*, Symp. Proc., 6th Commonwealth Mining & Metallurgical Congress, CIM, Ottawa, 1957, pp. 353-355.

156. Northern Miner, "Lorado's Mill Starts to Run," *The Northern Miner*, 1957, Apr. 4, p.8.

157. Northern Miner, "Commence Leaching at Lorado Uranium," *The Northern Miner*, 1957, Apr. 25, p. 21.

158. Northern Miner, "Lake Cinch Mines to Start Shipping in June to Lorado," *The Northern Miner*, 1957, May 9, p. 3.

159. Hicks, C.T.; Krekeler, J.H.; Nelli, J.R., "Laboratory and Pilot Plant Evaluation of Lorado Uranium Concentrate," Technical report NLCO 734, Office of Technical Services, Department of Commerce, Washington, D.C., 1958.

160. Eldorado, 1956 Annual Report, Eldorado Mining and Refining Ltd., Ottawa, 1957.

161. Eldorado, 1954 Annual Report, Eldorado Mining and Refining Ltd., Ottawa, 1955.

162. Eldorado, 1955 Annual Report, Eldorado Mining and Refining Ltd., Ottawa, 1956.

163. Burger, J.C.; Jardine, J.McN., "Canadian Refining Practice in the Production of Uranium Trioxide by Solvent Extraction with Tributyl Phosphate," Proc. 2nd United Nations Internat. Conf. on the Peaceful Uses of Atomic Energy, Geneva, June 1958, 17 pp.

164. Crawford, J.E., "Uranium and Radium," in *Minerals Yearbook, 1954*, U.S. Bureau of Mines, Washington, 1958, pp. 1241-1296.

165. World Nuclear Association, "How Uranium Ore is Made into Nuclear Fuel," World Nuclear Association, London, UK, 2016, http://www.world-nuclear.org/nuclear-basics/how-is-uranium-ore-made-into-nuclear-fuel.aspx.

166. Pitkanen, L.L., "A Hot Commodity: Uranium and Containment in the Nuclear State," Ph.D. Thesis, Department of Geography, University of Toronto, Toronto, 2014.

167. Globe and Mail, "New Uranium Refining Plan Being Installed by Eldorado," *The Globe and Mail*, **1954**, *Sept. 8*, p. 24.

168. Eldorado, "Annual Report 1958," Eldorado Mining and Refining Ltd., Ottawa, 1959.

169. Page, R.D.; Lane, A.D., "The Performance of Zirconium Alloy Clad UO$_2$ Fuel for Canadian Pressurized and Boiling Water Power Reactors," *Proc. Joint ANS-CNA Conference*, Toronto, June 10-12, 1968, Canadian Nuclear Association, Ottawa, Paper AECL-3068.

170. Bihun, G., Saskatchewan Ministry of Environment, *Personal communication* to SRC, May 2015.

171. Clifton Associates Ltd., "An Assessment of Abandoned Mines in Northern Saskatchewan (Year 2)," Report for Saskatchewan Environment, Regina, May 2002.

172. Golder Associates Ltd., "Technical Information Document for the Inactive Lorado Uranium Tailings Site," Report for EnCana Corp., April 2008.

173. Ruggles, R.G.; Robinson, D.J.; Zaidi, A., "A Study of Water Pollution in the Vicinity of Two Abandoned Uranium Mills in Northern Saskatchewan, 1978," Report EPS-MNR-5-81-2, Environment Canada, Ottawa, 1978.

174. Cooper, E.L.; Wagner, C.C., "The Effects of Acid Mine Drainage on Fish Populations," In: "Fish and Food Organisms in Acid Mine Waters of Pennsylvania," Report EPA-R3-73-032, U.S. Environmental Protection Agency, Washington, February 1973, pp. 75-124.

175. Joint Research Centre, "Advancing Implementation of Nuclear Decommissioning and Environmental Remediation Programmes," Policy Support Document EUR 27902, European Commission, Brussels, 2016.

176. Brown, L.D., "Proposed Decommissioning of the Gunnar and Lorado Uranium Mine Sites," Report for Saskatchewan Environment, BB Health Physics Services, Regina, SK, 1993.

177. Stenson, R.; Howard, D., "Regulatory Oversight of the Legacy Gunner Uranium Mine and Mill Site in Northern Saskatchewan, Canada – 13434," *Proc., Waste Management Conference (WM2013)*, WM Symposia, Inc., Tempe, AZ, 2013, 13 pp.

178. Peach, I.; Hovdebo, D., "Righting Past Wrongs: The Case for a Federal Role in Decommissioning and Reclaiming Abandoned Uranium Mines in Northern Saskatchewan," Public Policy Paper 21, Saskatchewan Institute of Public Policy, University of Regina, December 2003.

179. Saskatchewan, "Canada's New Government and Province of Saskatchewan Launch First Phase of Cleanup of Legacy Uranium Mines," News Release, Government of Saskatchewan, Regina, April 2, 2007.

180. Saskatchewan Research Council, "Risk Reduction Plan for the Inactive Lorado Uranium Tailings Site: Environmental Impact Statement," Saskatchewan Research Council, Saskatoon, November 2013.

181. SRC, "Project CLEANS (Cleanup of Abandoned Northern Sites)," Saskatchewan Research Council, Saskatoon, 2014, http://www.src.sk.ca/about/featured-projects/pages/project-cleans.aspx.

182. Editorial Board, "U.C. Meeting on Abandoned Mines," Supplement, *Opportunity North*, **2008**, *Spring*, 2-3.

183. Editorial Board, "Cleanup Process Begins at Gunnar," *Opportunity North*, **2009**, *Summer*, 22.

184. Editorial Board, "Project CLEANS Team Gears up for a New Work Season," *Opportunity North*, **2013**, *Spring*, 29 (see also p. 20).

185. Petelina, E., Sanscartier, D.; MacWilliam, S.; Ridsdale, R., "Environmental, Social, and Economic Benefits of Biochar Application for Land Reclamation Purposes," *Proc. 38th Ann. B.C. Mine Reclamation Symposium*, 2014, 13 pp.

186. Saskatchewan, "Northern Saskatchewan Environmental Quality Committee," Government of Saskatchewan, Regina, 2016, https://www.saskatchewan.ca/residents/first-nations-citizens/saskatchewan-first-nations-metis-and-northern-initiatives/northern-saskatchewan-environmental-quality-committee.

187. Saskatchewan Research Council, "Gunnar Site Remediation Project: Environmental Impact Statement," SRC 12194-320-1L13, February 2013.

188. Provost, K., "Abandoned Mine Being Cleaned-Up in Saskatchewan," CJLR-FM News, La Ronge, SK, broadcast 25 October 2010.

189. Muldoon, J.; Schramm, L.L., "Gunnar Uranium Mine Environmental Remediation – Northern Saskatchewan," Paper ICEM2009-16102, *Proc. 12th Internat. Conf. Environmental Remediation and Radioactive Waste Management - ICEM'09/DECOM'09*, Liverpool, U.K., October 11-15, 2009.

190. Muldoon, J.; Schramm, L.L., "Gunnar uranium mine remediation project. Northern Saskatchewan," *Proc. 33rd Arctic and Marine Oilspill Program (AMOP) Technical Seminar on Environmental Contamination and Response*, Halifax, N.S., June 7-9, pp. 383-403, 2010.

191. Wilson, I.; Allen, D.E.; Schramm, L.L.; Muldoon, J., "Lorado Uranium Mine Environmental Remediation – Northern Saskatchewan," *Proc. IAEA Internat. Conf. on Advancing the Global Implementation of Decommissioning and Environmental Remediation Programmes*, Madrid, Spain, May 23-27, 2016, paper IAEA-CN-238-30P.

192. Hawkins, T.R.; Beyersdorff, L.E.; Vitale, R.W., "Method of Stabilization and Dust Control," *U.S. Patent* 7,074,266, July 11, 2006.

193. Hawkins, T.R.; Beyersdorff, L.E.; Vitale, R.W., "Method of Chemical Soil Stabilization and Dust Control," *U.S. Patent* 7,081,270, July 25, 2006.

194. Midwest, "EK35® Synthetic Organic Dust Control," Material Safety Data Sheet (MSDS), Midwest Industrial Supply Inc., Canton, OH, June 2, 2011.

195. Jensen, S.; Sharp, T.; Allen, D.; Wilson I., "Design of In-Situ Water Treatment of Acid Contaminated Lake," *Proc. 38th Annual Mine Reclamation Symposium*, British Columbia Technical and Research Committee on Reclamation (TRCR), Prince George, BC, 2014.

196. Redmann, R.E.; Frankling, F.T., "Revegetation of Abandoned Uranium Mill Tailings Near Uranium City, Saskatchewan. Plant Species Selection," Report for Saskatchewan Environment, Regina, March 1982.
197. Larmour, A., "Elliot Lake Hailed as Reclamation Success Story," Sudbury Mining Solutions J., Sept. 1, 2010, http://www.sudburyminingsolutions.com/elliot-lake-hailed-as-reclamation-success-story.html.
198. Mawhiney, A-M.; Pitblado, J. (Eds.), *Boom Town Blues: Elliot Lake, Collapse and Revival in a Single Industry Community*, Dundurn Press, Toronto, 1999.
199. Waggitt, P., "Uranium Mining Legacy Sites and Remediation - A Global Perspective," Presented at IAEA Conference, Namibia, October 2007, International Atomic Energy Agency, http://www.iaea.org/OurWork/ST/NE/NEFW/documents/RawMate rials/CD_TM_Swakopmund%20200710/13%20Waggit4.PDF.
200. IAEA, "Advancing Decommissioning and Environmental Remediation Programmes," IAEA Nuclear Energy Series Report No. NW-T-1.10, International Atomic Energy Agency, Vienna, 2016.
201. CNSC, "Uranium Mines and Mills Waste," Canadian Nuclear Safety Commission, Ottawa, 2014, http://www.nuclearsafety.gc.ca/eng/waste/uranium-mines-and-millswaste/index.cfm.
202. AANDC, "Port Radium Mine (Remediation Complete)," Aboriginal Affairs and Northern Development Canada, Ottawa, 2012, http://www.aadnc-aandc.gc.ca/eng/1332423218253/1332441057035.
203. MacPherson, A. "Gunnar Cleanup to Exceed $250M, 10 Times Estimate," *Saskatoon StarPhoenix,* October 17, 2015, Last Updated: February 18, 2016.
204. MacPherson, A. "Overbudget Gunnar Cleanup Federal Responsibility, Sask. Politicians Say," *Saskatoon StarPhoenix*, February 24, 2016.
205. AAEC, "Rum Jungle Project," Booklet, Australian Atomic Energy Commission, Lucas Heights, Australia, 1963.
206. World Nuclear Association, "Former Australian Uranium Mines," World Nuclear Association, London, U.K., 2014, http://www.world-nuclear.org/info/Country-Profiles/Countries-A-F/ Appendices/Australia-s-former-uranium-mines.

Also of interest …

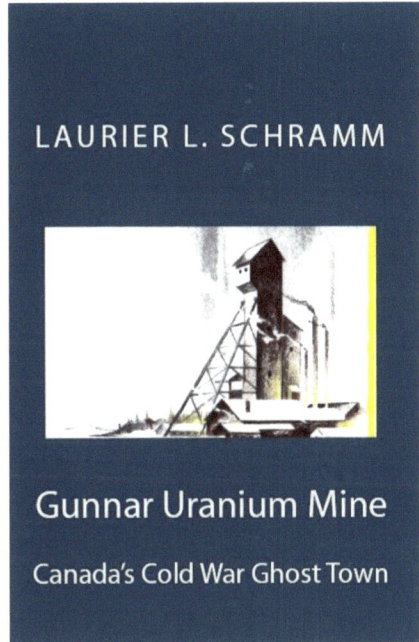

LAURIER L. SCHRAMM

Gunnar Uranium Mine

Canada's Cold War Ghost Town

The Gunnar mine, mill, and town-site were built in a remote location in northern Saskatchewan, on the shore of Lake Athabasca. Like most mining communities the town boomed, first with construction workers and miners, and later with families. When the Gunnar mill construction was completed in the fall of 1955 it doubled Canada's uranium production capacity. By 1956 the Gunnar mine was the largest uranium producer in the world. The Gunnar town-site was built to serve the mine and mill and at one time had a population of about 850 people. By 1964 it was a ghost town. The Gunnar mine produced over 5 million tonnes of uranium ore, nearly 4.4 million tonnes of mine tailings, and an estimated 2,710,700 m^3 of waste rock. Following closure in 1964, the Gunnar site was abandoned with little remediation and no reclamation being done. It has been referred to as "the second greatest environmental disaster area in Canada." Forty years would pass before the governments of Saskatchewan and Canada reached an agreement to fund the remediation (clean-up) of the Gunnar site, and contracted the management of the project to the Saskatchewan Research Council (SRC). At the time of writing this book, the clean-up was well underway, with several years of clean-up activity remaining, and a further expected 10-15 years of monitoring activity before the site is expected to be released into a long-term management and monitoring program.

Print ISBN: 978-0-9958081-2-6   **ePub ISBN:** 978-0-9958081-0-2

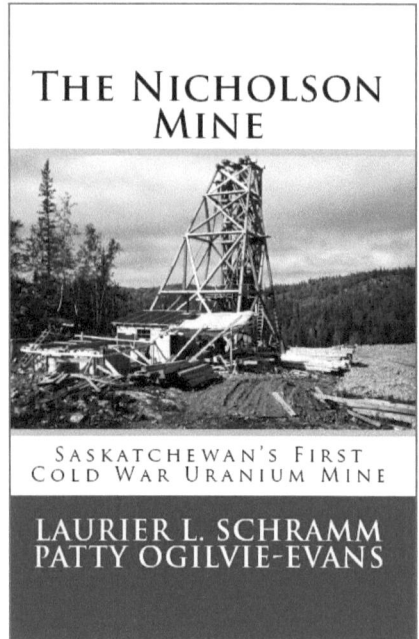

The first discovery of uranium in Saskatchewan was at Nicholson Bay, on the shore of Lake Athabasca. Uranium was first noted at what became the Nicholson site in 1929 when uranium was only of interest as an indicator of radium potential. When uranium ores became of strategic national interest in about 1940, a cross-Canada search was launched to find uranium deposits. The first to be found and developed was in the Northwest Territories. The second arose from a return to exploration at the Nicholson site in the Beaverlodge area in 1944. The Nicholson mine was the first uranium mine to be developed in Saskatchewan and, in 1949 was the only active uranium mine in Canada outside of the Northwest Territories. By 1959 the Nicholson ore body had been essentially depleted, but the Nicholson mine had played its role in helping Canada become one of the largest uranium producers in the world. It produced about 12,800 tonnes of uranium ore, yielding about 50 tonnes of uranium (as U_3O_8), and an estimated 60- to 90 thousand m^3 of waste rock. Following closure in 1960, the Nicholson site was abandoned with little remediation and no reclamation being done. Forty-five years would pass before the governments of Saskatchewan and Canada reached an agreement to fund the remediation (clean-up) of the Nicholson site, and contracted the management of the project to the Saskatchewan Research Council (SRC).

Print ISBN: 978-0-9958081-4-0   **ePub ISBN:** 978-0-9958081-5-7

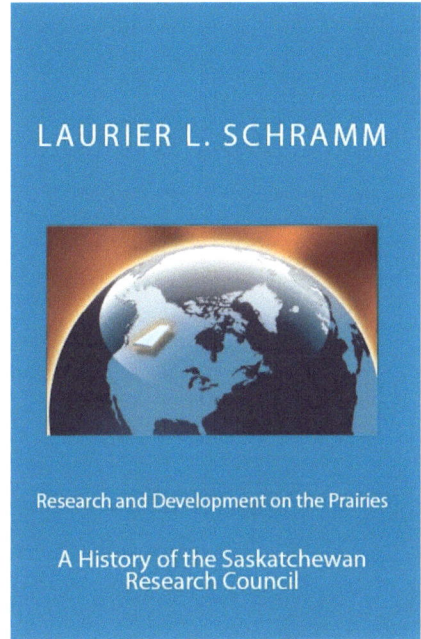

LAURIER L. SCHRAMM

Research and Development on the Prairies

A History of the Saskatchewan
Research Council

Early in the 20th century, the advent of industrial research councils brought organized research, development, and technological innovation to North America. Such organizations, now usually referred to as research and technology organizations (RTOs), focused on research and development aimed at helping industry develop and advance, and they were critical to the evolution of the modern approach to technological innovation. One of Canada's first RTOs to be established was the Saskatchewan Research Council (SRC), and it has become one of the most enduring. This book traces the evolution of SRC from its first efforts in the 1930s through several distinct eras.

Print ISBN: 978-0-9958081-3-3   **ePub ISBN: 978-0-9958081-1-9**

www.ingramcontent.com/pod-product-compliance
Lightning Source LLC
Chambersburg PA
CBHW050438240326
41599CB00060B/14